COCONUT PRODUCTION AND MARKETING

COCONUT PRODUCTION AND MARKETING

S. SURESH BABU

ABHIJEET PUBLICATIONS
DELHI 110094

ABHIJEET PUBLICATIONS
2/46 Tukhmeerpur Extension
Delhi-110 094
Phone: 011-22960492, 65698474
E-mail: abhijeetpublication@gmail.com

COCONUT PRODUCTION AND MARKETING

First Published 2010

ISBN 978-93-80031-49-1

PRINTED IN INDIA

Published by J.K. Singh for Abhijeet Publications, Delhi-110 094, Lasertypeset by Gaurav Graphics, New Delhi and Printed at Himanshu Printers, Delhi.

Preface

Agricultural production is highly influenced by nature on the one hand and on the other hand agricultural marketing is highly influenced by man-made arrangements. Coconut production and marketing is also has no special exception from this observable fact. The present status of area under coconut, production and productivity in India is so encouraging but the other side of the coin coconut marketing is imbibed with several unresolved problems. Indeed, the ultimate sufferers are poor Indian coconut growing community.

This book is a miniature attempt to explore the coconut production and marketing practices of coconut growers and their opinion towards the market functionaries. It also deals with production strategies, disease control, medicinal and industrial uses of coconut and it properly ends with suggesting some practicable measures to override the present problems.

It is impossible to write a book without the help and contributions of many other people. I owe a large debt to all my teachers and to authors of other coconut related books. I would like to single out a few academicians who have been direct assistance to me in the preparation of this book. My research supervisor Dr. P. Murugesan, Associate Professor, SRMV College of Arts and Science reviewed many of the original chapters with diligence and whose constructive criticism and suggestions helped invaluably in shaping the form and content of this text, and lovable colleague Dr. R. Karunakaran, Assistant Professor, Hawassa University who voluntarily extended his assistance in updating and revision of all the chapters.

I am indebted for many valuable suggestions made by my respectful teachers Dr. V. Kualandaiswamy, Principal Sri Saraswathi Thiyagaraja College, Dr. S. Nakkiran, Professor Ambo University, Dr. B. Subburaj, Professor and Head Gandhigram Rural University, and Dr. R. Ravichandran Associate Professor Gandhigram University.

I would like to dedicate this master piece of work to my parents R. Soundara Rajan Head Master (Rtd) and S. Banumathi for their continuous encouragement and moral support though out my education. Although it sounds formal, a great sense of acknowledgement goes to my wife J. Sumithra and my two heirs Master S. Sree Hari and Master S. Nithin for sparing their valuable time. My thanks are also due to V. Janarthanam Teacher (Rtd) and J. Selvarani for their moral support and motivation.

Words are inadequate to express my sincere thanks to Mr. J.K. Singh, publisher for having undertaken this assignment and the successful publication of my work.

Above all, I am grateful to God for His blessings and grace.

Last but not the least, although I appreciated all the excellent advice and counsel of others, errors and omissions that remain are the sole responsibility of the author.

Dr. S. Suresh Babu

Contents

Abbreviations

ADB	Asian Development Bank
AIDS	Acquired Immune Deficiency Syndrome
AOS	Alpha Olefin
APCC	Asian and Pacific Coconut Community
BIS	Bureau of Indian Standards
CBFS	Coconut Based Farming System
CDB	Coconut Development Board
CFTRI	Central Food Technological Research Institute
CGD	Chow Hat Green Dwarf
CMS	Cooperative Marketing Society
COD	Chow Hat Orange Dwarf
CPCRI	Central Plantation Crops Research Institute
CSIR	Council for Scientific and Industrial Research
D × D	Dwarf × Dwarf
D × T	Dwarf × Tall
DAC	Department of Agriculture and Cooperation
DC	Desiccated Amount
DFRL	Defense Food Research Laboratory
ECT	East Coast Tall
FGD	Focus Group Discussion
FAO	Food and Agricultural Organization
gm.	Gms
GOI	Government of India
ha.	Hectare

HDL	High Density Lipoprotein
HIV	Human Immune Virus
ICAR	Indian Council of Agricultural Research
ICCC	Indian Central Coconut Committee
IPGRA	International Plant Genetic Resources Institute
KERAFED	Kerala Coconut Growers' Federation
kgs.	Kilograms
LABS	Linear Alkyl Benzene Sulphonate
LDL	Low Density Lipoprotein
lit.	Litre
MARKFTFED	Kerala State Cooperative Marketing Federation
MGD	Malayan Green Dwarf
MOD	Malayan Orange Dwarf
ml.	Milli Litre
MYD	Malayan Yellow Dwarf
NAFED	National Cooperative Agricultural Marketing Federation
NBF	Non-Banking Financial Institutions
PACB	Primary Agricultural Cooperative Bank
PCARDB	Primary Cooperative Agricultural and Rural Development Bank
PSA	Preference Score Analysis
RRL	Regional Research Laboratory
Rs.	Rupees (Indian)
SAU	State Agricultural Universities
SIS	Structured Interview Schedule
SPSS	Statistical Package for Social Sciences
sq.m.	Square metre
TANFED	Tamil Nadu Cooperative Marketing Federation
TCS	Water of Tender Coconut
T × D	Tall × Dwarf
T × T	Tall × Tall
TMOC	Technology Mission On Coconut
TNAU	Tamil Nadu Agricultural University
VCDC	Vayalar Community Development Centre
WCT	West Coast Tall

List of Tables

1

Design and Execution of the Study

Introduction

Coconut, the versatile palm popularly known as the 'Tree of Life', Tree of Heaven',' Tree of Abundance', 'Nature's Super Market', 'King of Palms', 'Kalpavriksha' as well as 'God's Gift to Mankind' and its fruit as *Lakshmiphal,* the Fruit of Wealth. It is grown in more than 93 countries (P. Rethinam, 2002). Its kernel, water, shell, husk, leaf, trunk, oil, flowers etc., have a wide variety of use. It provides food, drink, health, medicine, shelter, aesthetic materials and wealth for millions of people in the Asian and Pacific region by making use of its products and byproducts. Coconut is a crop of great antiquity in India. It is a traditional plantation crop grown in India for the last 3000 years and thus possesses the longest recorded history in the country. The crop has significant role on the national economy besides its influence on the economic, social and cultural lives of millions of small and marginal farmers who form the backbone of the coconut culture and industry of the country. It acts as an enormous source of raw material for various small, medium and large-scale industries, particularly in those states where this crop is largely grown.

The coconut palm, *Cocos nucifera* is found along the coast and in the interior of almost all tropical countries between the tropics of Cancer and Capricorn. Its wide distribution has been favoured by its usefulness as well as by its adaptability to different ecological conditions. In many tropical countries, coconut is an important part of the daily diet. The oil extracted from the kernel is the main coconut product. The residue is an important animal feed. The coconut water from the young nuts is healthy and nutritive beverage. The jelly-like kernel of the young coconut is considered a delicacy. The shredded and dried kernel known as desiccated coconut is used in food and confectionary. The husk of the nut provides an important fiber that can be used for ropes, carpets, brushes, geo textiles etc. The shell of the nut is used for household utensils, handicraft items and the charcoal made from it is an excellent basic material for activated carbon. Instead of being used for nut production, the inflorescences can be tapped, yielding sap with high sugar content, from which sugar, alcoholic beverages and vinegar can be made. The leaves are used for roof thatching. The mid-ribs of the leaves are used for brooms. Coconut wood is being used for house building, furniture items and tool handles. It is no wonder that this tree is called 'Tree of life' and Kalpavriksha.

Origin of Coconut

The country of origin of the coconut is unknown and various regions have been indicated by various scientists, from South America to Melanesia, Asia and Madagascar (Ohler 1984). Although the original home of the coconut is still unknown, among the theories which have been developed so far, two have been disputed for a long time. One locates the origin of the coconut palm somewhere at the northern end of the Andes in South America; the other indicated Southeast Asia as its home country. The coconut (Menon and Pandalai, 1958) was introduced to India in the post-Vedic period and in spite of the discovery of fossil Cocos species in Rajasthan, India cannot be considered the original home of coconut. According to geographical history, Rajasthan desert was once covered by the sea and it is possible that remnants of trees could have been brought down by oceanic currents. However, if tree remnants can be brought down by sea currents, so can coconuts. The theory of the later import in India is supported by the fact that among the names of the coconut in Malaysia, those of Sanskrit derivation were of a later origin the others already prevalent in the area. Menon and Pandalai concluded that coconuts originated somewhere in what is now called Melanesia.

In India the first recorded history of coconut dates back to Valmiki *Ramayana* period. In the Valmiki *Ramayana* there are references to coconut in Kishkindha Kanda and Aranya Kanda. Valmiki *Ramayana* dates back to probably the 3rd Century BC and hence it can be safely presumed that coconut was introduced to India *during the Post-Vedic period* (Nair 1996).

Statement of the Problem

Marketing is as important as production to any producer. Because it creates value to the product; it pays revenue to the producer; and more than these, it directs the producer as to whether continue or stop production. To a farmer, marketing is something more than production, due to certain inherent features that neither the production can be controlled/regulated in tune to market changes due to predominance of natural forces affecting production functions nor can marketing be performed in tune to market requirements due to his own internal constraints. Hence farmers remain with chronic problem that they can neither derive the advantages of their production/productivity increase nor the advantages of better marketing. It is apt to remember the findings of Dantwala Commission on Cooperative Marketing (1966) and FAO (1988) that agricultural production is abundant in our country while the problem is with distribution (Marketing). In all the years farmers were deprived of the benefits of increased production in terms of market price and vice versa. Equalization of demand and supply functions in agricultural marketing is far from any body guess. Coconut farmers are not exempted from the above phenomenon.

Marketing determines the productivity and profitability. Creating a sound marketing system is essential, as the problems of coconut growers mostly emerge due to the imperfections in the marketing system. Sincere attempts have been made to clear-off the marketing imperfections especially after independence during plan periods. Passing of Agricultural Produces Marketing Act, organization of Regulated Markets, and Cooperative Marketing Societies and establishment of Ware Housing Corporations are important landmarks in this regard. Nevertheless there are group projects of coconut farmers as effective institutions have been found in some of the regions of Tamil Nadu. Such group projects are also observed in some developing countries like Taiwan, Thailand and Philippines (Aurora Regaldo, 1994). These group projects enhance collective bargaining among coconut growers and enable for providing competitive price to coconut farmers. They seem to function as bargaining cooperatives of Canada. Thus, development of different forms of marketing institutions questions the relevance of the conventional formal marketing structures. Non-Institutional Agencies still dominate in coconut marketing. If marketing is imbibed with imperfections, coconut farmers should have received low returns. This in turn could have resulted for reduction in coconut production. On the contrary, it is amazing to notice that both the area under coconut cultivation and production of coconut have been increasing year after year. The foregoing issue motivates the researcher to conduct a new research study in the sample area and to find out the answers to the following basic questions. Why do farmers adopt cultivation of coconut? What is the extent of adoption of coconut palms by them? Where and how do they dispose their coconut and coconut products? Do farmers undertake value addition for their coconuts? What are the channels of distribution for coconuts? Who are the buyers? What is the role of institutional marketing bodies like Cooperative Marketing Societies and Regulated Markets in coconut marketing? What is the role of Non – Institutional Agencies like Traders, Commission Agents, Brokers, Itinerant Traders and Industrial User and Wholesalers in coconut marketing? Do they serve for better marketing of coconuts and coconut products by coconut growers? What is the perception of farmers on the marketing services of Institutional and Non-institutional Agencies? How do the farmers overcome their marketing problems? What do farmers suggest for solving the marketing problems? To explore the answers to the above issues, an in-depth empirical analysis is a sine-qua-non.

Objectives of the Study

The prime objectives of the study are:

i) To assess the coconut production at macro and micro level
ii) To analyze the production practices of coconut growers
iii) To analyze the existing marketing practices of coconut growers and examine the channels of distribution of coconuts in the study area
iv) To evaluate the perception of coconut growers about the present coconut marketing functionaries.

Hypotheses

Based on the above objectives, the following hypotheses have been formulated for the study:

a) There is no significant relationship between educational qualification of the respondents and variety of coconut tree cultivated
b) There is no significant relationship between marketing channels and price offered by them

c) There is no significant difference between organic input and inorganic input in the average coconut sales per acre
d) There is no significant difference between single crop farming group and mixed crop farming group in the average quantity produced per acre.
e) There is no significant difference among the type of farmer groups in the average quantity produced per acre.
f) There is no significant difference among the type of farmer groups in the average of preference score towards traders.

Methodology

The study is an empirical analysis of coconut production and marketing practices in Tamil Nadu with special reference to Coimbatore District. The data both qualitative and quantitative were gathered through Field Survey Method and Personal Interview Technique. The secondary data were collected from the journals, reports, books, internet etc.

Sampling Procedure

A Multi-Stage sampling procedure was adopted for the present study

Stage-I: Selection of District

All the districts in Tamil Nadu were arranged in descending order on the basis of their area and production under coconut. Since the Coimbatore district is a leading coconut growing area in Tamil Nadu, it was purposively selected (See Appendix – I).

Stage-II: Selection of Block

At the second stage, all the blocks in the Coimbatore district were arranged in descending order on the basis of their area and production under coconut. The Anamalai Block had the largest area and production under coconut, So it was considered and purposively selected (See Appendix – II).

Stage – III: Selection of Coconut Growing Villages

At the third stage, all the villages in the Anamalai Block were arranged in descending order on the basis of area and production under coconut. The villages found under the first quartile, the second quartile and the third quartile were considered High, Moderate and Low category respectively in respect of coconut crop. Two villages from each category of villages were selected by adopting simple random sampling procedure. Thus the following six villages were selected for the study i) Vettaikaranpudur ii) Marchinaickenpalayam iii) Jallipatti iv) Anamalai v) Thensengampalayam and vi) Thensittur (See Appendix – III).

Stage-IV: Selection of Respondents

51 farmers each from six villages totaling 306 respondents were selected for the study by adopting Quota Sampling procedure.

Tools and Techniques Used

Field Survey, Interview schedule, Focused Group Discussions and Informal discussions are planned for eliciting data and information's from the coconut growers and institutions.

The interview method is used to collect information from the sample farmers.

Focused Group Discussion and Informal Discussions are used to collect information with the prime informants such as Coconut farmers, Traders, Commission Agents, Brokers, wholesalers and Cooperative Marketing Society officials and leaders, officials of Regulated Markets, Government Officials etc.,

Field observations are made from the farmer's field, wholesale and retail market yards, rural and urban market centers, commission mandies, broker's officer, cooperative marketing societies, regulated markets, production and processing centers etc.,

By adopting the above tools and techniques the information regarding personal particulars, size of the farm, area under coconut and other crops, asset position, cropping pattern, particulars about coconut gardens, reasons for preferring coconut farming, establishment and maintenance cost of the garden, yield pattern of coconuts, income from the coconuts, problems in the coconut cultivation, marketing practices, opinion about existing system of marketing, role of institutional agencies, problems with the institutional agencies, advantages in the institutional agencies, role of Non-institutional agencies and its merits and demerits, marketing practices followed by the non-institutional agencies, perception about the institutional and non-institutional agencies etc., are collected.

Data Analysis and Interpretation

Data obtained from the field were analyzed with the help of Statistical Package for Social Sciences (SPSS) latest version. The survey statements were statistically tested by employing statistical tools such as Averages, Range, Standard Deviation, ANOVA, Chi square test, Paired Sample Test and t. test.

To understand perception of the farmers towards the present market functionaries Preference Score Analysis (PSA) was employed.

Scope of the Study

The present study aims at assessing the coconut production at macro and micro level, analyzing the existing marketing practices and problems of coconut growers and examine the channels of distribution of coconuts in the study area and estimating the share of Institutional and Non- Institutional agencies in coconut marketing and evaluating the perception of coconut growers about the present scenario of coconut marketing.

Limitations of the Study

The study was confined to Anamalai Block in Coimbatore District, which is unique in agro-climatic condition, and the findings of the study may be used with caution, one could understand the problems in collecting and analyzing the data relating to perennial crops. Since the farmers are not in the habit of recording the data regarding to the coconut yield, marketed price, income and expenses, cross checks were made in eliciting the information to reduce the recall bias. It is obvious that the findings of the study have their own limitations in their applicability to other parts.

Delimitations of the Study

The study is also delimited mainly to the marketing of coconuts and to a major assembling marketing area. It has not included marketing of coconut by-products within its purview. Despite the limitations, best effort was extended to make the study a qualitative one

Organisation of the Report

The report is organised into seven chapters:

- The first chapter outlines the Design and Execution of the Study
- The second chapter presents Review of Literature
- The third chapter gives a bird's eye view on Historical Perspectives of Coconut Farming.
- The fourth chapter narrates Coconut Marketing: practices and problems in study area.
- The fifth chapter describes the production practices of coconut growers
- The sixth chapter expresses coconut growers' perception about the present marketing functionaries
- The final and seventh chapter presents findings, conclusions and suggestions.

2

Review of Literature

Several research studies have been conducted in several parts of India during the past few decades related to coconut growing area, production, productivity, market functionaries, markets of coconut based products and byproducts, state and central government support and coconut based non-farm sector activities. Perusal of the related literature shows that only a very few research studies have taken in assessing the role of institutional and non-institutional agencies in coconut marketing. Some of the important studies relevant to the present research are reviewed here.

Geoffery, Bastin (1986) conducted a study on "Coconut marketing: issues for a threatened industry". He reviewed market conditions in the laurie sub-sector, describes the marketing channels for copra and coconut oil and sets out detail of significant issues for the future. It concluded that the industry must take urgent steps to improve the flow of market information, develop cost saving means of production which lead to a range of items and investigate new markets for non-traditional products.

Raveendran, P. (1987) in his study on "Demand, consumption pattern and consumer acceptability of desiccated coconut" in Karnataka viewed that though the first desiccated coconut unit was established in the year 1952, the industry is still in its infancy. There are at present 26 such units in the country, Karnataka is the leading producer of desiccated coconut with about 18 units accounting for more than 75 percent of the total production in the country. The present total annual production is estimated at about 10,000 to 12,000 tonnes. This industry provides employment to more than 1500 people on a regular basis.

Raveendran, P. (1989) conducted a study on Marketing of coconuts and its products in Andaman and Nichobar Islands with the main objectives of trends in coconut production, to educate the farmers for adopting modern agricultural practices, and developing a good market for coconuts. The study came to the conclusion that the cooperative marketing societies at all levels need to be further strengthened. The state and central governments may extend all financial and other helps such as provision for soft loan to apex societies for outright purchase and provision for providing transport facilities at concessional rates.

Ratha Krishnan, L. (1989) made an investigation to analyze the "Price spread in coconut marketing" in the village M. Vadipatty, Anna District, Tamil Nadu. The prime objective was to study the price spread of coconut among small coconut growers. He analyzed the present position of marketing, market functionaries and price spread of coconuts among the existing channels. He found that coconut is transferred from the producer to the final consumer through four channels they are: *i)* primary producer to final consumer, *ii)* Primary producer to wholesaler to retailer to final consumer, *iii)* primary producer to retailer to final consumer, and *iv)* Primary producer to middlemen to wholesaler to retailer to final consumer. No price spread is found in channel one. Gross and net prices spread are greater in channel three. Price spread could be reduced either through farmers cooperative marketing pattern or through the government intervention.

Hameed Khan, H. and others (1990) in their study on "Improving the coconut production" in Kerala expressed that 4500 to 5000 nuts are required to produce one tonne of copra at the international scenario, corresponding figure for India would be 6800 nuts. The per capita availability of coconut in the country is as low as 10 nuts/year where as it is high as 53 nuts in Indonesia and 124 nuts in Sri Lanka.

Punchihewa, P.G. (1990) conducted a study on, "Coconut industry—current situation and prospects" in Jakarta, Indonesia. He found that coconut is the most extensively grown nut in the world and the most important of palms. It is also found that in nearly ninety countries the annual production ranging of 12 billion nuts in Philippines to 2.5 million nuts in Nauru. On the basis of the average production for the period 1984-1988 the present acreage is estimated to nearly 41 billion nuts a year.

Anitha Sharma (1990) found her study on "Villagers can do well with better marketing" that the green revolution has made the village marketing boom. The consequence has been that producers are being subjected to exploitation. Further the cooperatives too have not been able to deliver the goods and the middlemen rule supreme in village markets. The old marketing system has been going out of use. This has created a gap, which needs to be filled.

George, M.V., and others (1991) conducted a study on "Trend in area production and productivity of coconut in India" with prime objective of understanding trend in area, production and productivity of coconut. The main findings of this study were the area under coconut in Kerala and Karnataka recorded a steady growth. In Tamil Nadu though there was a reduction in the area under coconut during fifties whereas steep increase in the area was seen especially during earlier sixties and during the second half of eighties. The total production of coconut during 1950-51was only 3582 million nuts which rose to 9283 million nuts during 1989-90 recorded an increase of 157 percent over a period of 40 years. The average productivity in 1950-51 was 5259 million nuts that rose to 6150 million nuts in 1989-90. India ranks third among the coconut growing countries. The state of Kerala which accounts for about one percent of the total land area of India, contributes to about 56.4 percent of the area and 47 percent of the production of coconut. The other states viz., Tamil Nadu, Karnataka and Andhra Pradesh together account for 34 percent of area but contribute 44 percent of the production. All the other states together, account for 9 percent of area and production of coconut in India.

Thomas, Mathew (1991) conducted a study on, "The Consumption of tender coconut in Tamil Nadu" with the objective to assess the actual consumption and contribution of this trade to the state's economy. The result of the study indicated that the consumption of tender coconuts in the state is increasing unprecedently. The average consumption of tender coconuts in a city in Tamil Nadu was estimated at 23,063 lakhs and that of the state was 122.223 million nuts per annum. The demand for the tender nuts are high in summer months and the average price was Rs. 2000/- per thousand nuts. The total sales added to the state's economy by way of sales of tender coconuts were worked out at Rs. 24.45 crores. The trade of tender coconuts provided employment opportunities to agricultural labourers, during the off-season. The trend in the consumption of tender coconut indicated that bottled coconut water could be introduced as a soft drink in important cities in the state.

Jaganathan, N. (1991) made an attempt in this study on "An

Economic Analysis of Coconut Farming in Anaimalai Block of Coimbatore District of Tamil Nadu" to examine cost structure, resource use efficiency in terms of coconut productivity and productivity of capital investment of coconut farming. The preliminary survey schedule was tested and the primary data were collected based on the revised schedule. He used appropriate statistical tools to select 109 coconut farmers. The study reveals that farmers shifted their cropping pattern towards coconut farming since coconut farming with intercrops gives more benefit-cost ratio through more resource use efficiency. The functional analysis indicates that number of bearing trees, organic manures and irrigation contributed more for yield. The coconut complexes can be organized in the producing centers with a net work of small scale industrial units like copra milling, manufacture of coir and coir products, mat making and other related industries which could generate additional employment and make use of the increased availability of coconuts and by-products.

Ron Harris (1991) in his paper "The coconut industry utilization and marketing" concentrates on examination of the international market for desiccated coconut. He reveals that Philippines stands out as the major supplying nation, and accounts for well over half of total world exports. The second largest producer is Sri Lanka, which currently supplies approximately one third of the world export market. Large-scale production in the Philippines has lent itself to easy bulk handling, via containerization. The channel of desiccated coconut trade is further simplified by historical links between manufacturers in Philippines and importing companies in USA.

Jos, C.A. (1992) conducted a study on Cooperativisation of marketing and processing of coconuts, problems and prospects in Kerala . The main objectives of the study were to identify the malpractices involved in the existing marketing system and to organise the small farmers for selling their produce at a profitable bargain. The findings of the study reveal that the education at the gross root level will be the primary task of organizing the coconut producers for group action. This may be done by trained specialists called spearhead team as in the case of dairy projects of National Dairy Development Board. The role of spearhead team falls into four categories: *a)* Entrepreneurial, *b)* Rural communication, *c)* Managerial and *d)* Training.

Prafulla K. Das (1992) conducted a study on Coconut marketing problems, prospects and challenges in Kerala with the objectives to identify specific problems in coconut marketing and processing sectors, to find out various issues in connection with those problems and to suggest appropriate measures for solving those problems. The study revealed that 81 percent of marketed surplus nuts in Kerala are sold to the middlemen by the farmers at their farm gates, only 17.5 percent of marketed surplus find through market outlets and remaining 1.5 percent are disposed of through the cooperative network. He list outs the problem like middlemen abuse, lack of facilities for copra making, lack of proper copra processing facilities, price instability etc.

Ramakrishna Pillai, K. (1992) conducted a study on role of cooperatives in processing and marketing of coconuts in Kerala with special reference to Kerala State Cooperative Marketing Federation. The main objective of the study deals with the role of Kerala State Cooperative Marketing Federation in processing and marketing of coconuts. The study concluded that the cooperative marketing is the only way to provide an institutional arrangement to solve the marketing problems of the farmers. The cooperative sector has to compete with the private sector and public sector. The effectiveness of the cooperatives in marketing will depend upon their efficiency productivity and services.

Markose, V.T. and others (1993) in their paper on "Coconut based rural industries – an overall perspective" stated that the coconut industry in India sustains roughly 10 million people in cultivation, processing, trade, transportation and other related activities. The total value of production of the crop is around Rs. 4000 crores and foreign exchange is around Rs. 100 crores through export of coconut-based products. Coconut processing sector in India is currently confined to copra production, oil extraction, coir manufacture, desiccated coconut manufacture and toddy tapping. The study reveals that in India as much as 60 percent of the coconut production estimated to be used for edible and religious purposes, 3.5 percent as tender coconut, 35 percent as milling copra for oil extraction and the balance is processed into products like desiccated coconut and coconut milk.

Chengappa, P.G. and others (1993) conducted a study on, "Methods of disposal of coconut and economics of processing coconut at farm level" with the aim to understand the pattern of disposal of coconut at the farm level and economics of copra making in the main coconut growing regions of Karnataka. The study reveals that preparation of copra and then selling it is more profitable than selling coconut as fresh nuts. However, while deciding copra manufacturing, the risk in terms of price fluctuations during storage period and availability of storage facility also needs careful consideration.

Markose, V.T. (1994) expressed in his study on, "Coconut development in the last decade" that more than 10 million people depend directly or indirectly on the coconut crop for their livelihood by its cultivation, processing trade and industrial activities. The coir industry, which provides direct employment to more than 4 lakh workers, derives it raw material from coconut. Export of coir and coir products earns foreign exchange to the tune of Rs. 100 crores every year. The area under and production of coconut as per the latest estimate available is 1.518 million hectares and 10 billion nuts respectively.

Marapandiyan, P. (1994) stated in his study on, "Role of KERAFED (Kerala Coconut Growers' Federation) in coconut processing and marketing" that KERAFED has a built-in system to procure copra on a regular basis through the member Primary Agricultural Cooperative Society. In addition, whenever there were crisis times it had responded by intervening in the market. In 1989 and 1990, under the support price operations for copra declared by the Government of India, KERAFED was authorized as the representatives of NAFED to procure copra in the state.

Narasimhappa (1995) conducted a study on "Processing and marketing of coconut and its products" in Kerala with the aim of studying present arrangements in processing of coconuts, present system of marketing and problems encountered. The study revealed that although coconut is one of the most priced and essential commodities in India, its marketing is jeopardized by many artificial reasons and faulty Governmental policies and efforts are also needed to revamp the present unscientific marketing system in favour of a grower-consumer linked system. By involving growers in processing-marketing activities vertical integration could be achieved.

Haridas, R. and Chandran, C. (1995) conducted a study on Marketing systems, costs, margin, price spread and marketing problems of coconut—A case study of coconut growers and traders in Tamil Nadu. In this study they made an attempt to study the marketing system in terms of marketing practices such as harvesting grading, packing, marketing channels and to study the marketing costs, margins, price spread, and effects of the variations in the consumer price on the shares of the producer–seller and the retailer and the efficiency of marketing. The important findings of the study were: *a)* The share of producer in the net retail price of Rs. 3,015.18 per 1000 coconuts. b) The marketing margin to the wholesalers per 1000 coconuts is found to be Rs.170 and the retailers share is 265.28. c) Among the problems in coconut marketing lack of finance comes first in ranking followed by lack of transport facilities and storage facilities.

Aravindakshan, M. (1995) in his paper on "Challenges to the coconut industry in india and strategy for making it competitive" felt that India is producing 12,355 million nuts from an area of 1.63 million ha. The corresponding figure a decade back was 5807 million nuts and 1.16 million ha. The productivity level at present recorded in 7572 nuts per ha which was as low as 4982 nuts per ha in 1983-84. Increase in production of coconut has posed serious challenges to the coconut industry. At present coconut industry in the country is mainly centered around coconut oil alone. The large quantity of raw nut consumption to the extent of 60 percent of the total production leaves very little for marketable surplus or exportable surplus. Intensive efforts on product diversification and byproduct utilization are the immediate necessity to make the Indian industry remunerative and globally competitive.

Thampan, P.K. (1996) stated in his study on, "Coconut diversification, a must" that coconut production in India has been increasing over the years. From the third place on the world coconut map India has emerged second relegating the Philippines to the third position. In 1994 its share in global production was 24.1 percent and in area 14.7 percent. He said that the stability of coconut industry still hovers around a single product, coconut oil, which has already become vulnerable to substitution in all the end uses where it was once considered indispensable. Even with a viable market promotion programme for coconut oil, the stability of the industry cannot be sustained unless concerned attempts are made to utilize the multiple products offered by the coconut palm. This approach is essential to free the coconut based economy of the producing states from dependence on the fortunes of coconut oil alone. In product diversification and by-product utilization, facilities should be created for technological research pilot testing and entrepreneurial development. Effective marketing strategies have to be evolved for sustaining consumer demand for the new products. Greater attention had to be devoted in developing such products that can be competitively marketed in export markets.

Thampan, P.K. and others (1996) expressed in their study

on, "Global coconut situation and strategy for development" that the coconut production the world over from small and marginal holdings. These holdings range in size from 0.2 to 4 ha and account for over 95 percent of the total output. The annual production of coconut oil from these holdings accounts for less than four percent of the global output of major vegetable oils. For the period 1985 to 1994 Thailand recorded the maximum growth rate of 8.1 percent. The corresponding growth rates were 6.7 percent in India, 3.6 percent in Indonesia and 1.2 percent in the Philippines.

Thomas, P.T. and others (1997) stated in their study on, "Investment opportunities in coconut in Tamil Nadu" that among the major coconut growing states in India, Tamil Nadu occupies the second position after Kerala, sharing 33 percent of the country's coconut production. Tamil Nadu has made a spectacular increase from 15 percent to 33 percent to all India production of coconut during the last ten years. The state has recorded significant growth rate of 7.8 percent in area and 11 percent in production during the last decade. The productivity of coconut during the same period has increased from 10.988 nuts/ha to 14,554 nut/ha. Production of coconut would in all probability double in the next five years to 8700 million nuts resulting in a marketable surplus of about 3000 million nuts.

Venkattakumar, R. (1997) conducted a study on Socio-economic analysis of commercial coconut growers with the aim to find out the motivational factors influencing coconut cultivation, the knowledge level and extent of adoption of commercial coconut growers and the relationship of their characteristics with the knowledge level and extent of adoption, the socio-economic changes as a result of coconut cultivation, constraints in coconut cultivation and suggestion to overcome these constraints. The study revealed that most of the commercial coconut growers were motivated by the influencing factors viz., coconut is perennial crop, no need for intensive supervision, easy to do intercultural operation, less labour intensiveness, less cost of cultivation, more price for the produce in the market.

Subburaj, B. (2000) conducted a Research study on marketing of coconuts with a view to know the strategies employed for disposal of coconuts by farmers and the channel functionaries, seasonal variations in sale practices, mode of disposal of coconuts, indirect channel functionaries and their procurement and sale practices. This study is empirical analysis. It employed participatory appraisal technique. The study concludes that non-availability of organized credit for production, marketing and consumption purposes, absence of organized marketing institutions and difficulties in the adoption of off-farm value addition, processing and marketing technologies on coconuts are the major reasons for the 'forced sale' either through lease practices or on-farm sale of coconuts to marketing intermediaries.

Rethinam, P. (2001) conducted a study on problems and prospects of coconut industry in Andaman Nicobar Islands with a aim of studying coconut situation in Andaman & Nicobar Islands, trend in area, production and productivity of coconut, processing industry, prevailing marketing system and consumption pattern of coconut. The study reveals that Andaman & Nicobar Islands contribute an area of 24796 ha under coconut cultivation. The total palm population is 44.584 lakh and the annual production is 87.5 million nuts. Per ha productivity is as low as 3536 nuts per ha against the national average productivity of 7821 nuts per ha. This island recorded the lowest productivity rate in the country. Of the total production of coconut 40 percent is processed and the rest is consumed for house hold edible uses and tender nut purpose. The marketing system in this island was in vogue till recently wherein the tribes exchanged coconut for their daily requirements. The prevailing market price of copra in this island is only around Rs. 12-15 per kg. The Central Tribal Cooperative Society Ltd. does the procurement under the price support scheme. The marketing structure of coconut in this island is a two tier system i.e., primary society and central society under cooperatives.

Nair, M.K. (2001) conducted a study on coconut production and productivity with the objective to study the present status of coconut production in the world, production and productivity in different states of India and factors responsible for increasing production and productivity. The study reveals that the world coconut production (1999) is estimated at 54.129 million nuts from an area of about 12 million hectares. Nearly three-fourth of the production (73%) is contributed by India, Indonesia and Philippines. Sri lanka with about 5.22 percent of the production occupies fourth position. Kerala, Tamil Nadu, Andhra Pradesh and Karnataka are the major coconut producing states in India and together accounts for more than 80 percent of area and production in the country. Among the four states, Kerala accounts for largest area and production sharing 56.50 percent of area and 44.70 percent of production followed by Karnataka with 15.08 percent area and Tamil Nadu with 20.75 percent of production. Among the four states, Andhra Pradesh tops in productivity with 19.571 nuts per ha followed by Tamil Nadu with 11620 nuts per ha. Karnataka has the lowest productivity of 5195 nuts per ha. Kerala, which contributes to maximum production, has a productivity of 6188 nuts per ha. The maximum productivity of 15020 nuts per ha has been reported from Maharastra though it shares to overall production in the country is only 1.5 percent.

Rethinam, P. and Thampan, P.K. (2001) conducted a study on Reviving coconut based economy of Kerala state with the objective to create opportunities for enhanced on-farm income and employment, to promote efficient product and bye-product utilization both at the on-farm and community levels, to strengthen marketing infrastructure for domestic and export marketing and to direct research on varietal improvement for higher output of primary products from coconut and technology development. He suggested the following recommendation for the improvement of coconut based economy of Kerala. Coconut farmers in the state should be encouraged to organise themselves into self help groups or cooperatives for practicing intensive integrated farming to create multiple sources of income, food and employment. Support should be extended to farmers groups for efficient marketing of the multiple products obtained from integrated farming units both in the fresh and processed forms. Women groups should be organised for efficient product utilization both at the farm-household and community levels. Coconut based handicrafts should be encouraged by providing training support in design and techno-economic support in establishing production units and creating marketing infrastructure to the local artisans including women groups.

NABARD (2002) conducted a study on, "Coconut development in Tamil Nadu—An ex-post evaluation study in Coimbatore District" with the objective of studying the socio-economic profile of borrowers, to examine the scheme formulation and their implementation, financial viability of coconut farmers. The study revealed that the average price of nut during the reference year was Rs.2.25. Income received from the coconut sale was Rs.3.035 per acre and Rs.20,124 per farm during fifth year of investment. Repayment performance by the sample during the first two years after investment registered more than 95 percent.

Thomas Mathew, M. (2002) conducted a study on Trade in Tender coconut—A future vista of Indian coconut industry. He attempted to know the market behaviour of tender markets in the Maddur market in Karnataka state and he analyses its impact in inducing competitiveness and marketing abilities among the farmers.

The study revealed that the trend in the arrivals of tender coconut in the market during the last decade is increasing at a compound growth rate of 19 percent per annum. The seasonal indices for prices and arrivals indicated intra-year seasonal variations. The tender coconut price showed buoyancy during the summer months. Similarly the study showed an inverse relationship between the price and arrivals of tender coconut. Further the price of tender coconut in Karnataka moves in chose sympathy with the lagged price of the ball copra and coconut. The study draws an inference that the promotion of tender coconut in the country is helpful in price stabilization of coconut in the country.

Singh, R.K. and Subburaj, B. (2002) conducted a study on Pricing the coconut in Tamil Nadu—An Analysis with an attempt to address a few issues: What is the gross farm price for coconuts and what are the factors affecting net farm price for coconuts in Tamil Nadu? What is the trend in coconut prices in different market centres in Tamil Nadu and how do market price for coconuts determined? What are the factors influencing fixing of prices for coconuts at micro and macro level? The study revealed that the price realization through direct channel is higher than that through indirect channel. The absolute gross price spread is low in direct channel. However, the net price realized through local traders/commission agents/traders is the least due to high cost of marketing. Yet the price realized through the regulated market is the highest. The coconut price records a fluctuating trend throughout the year. The lowest price prevailed in March and September. Thereafter the price tends to increase till December, the highest being in December. He concluded that neither farmers nor traders have say in fixing the coconut price in Tamil Nadu.

He concludes that indirect mode of disposal of coconuts as a strategy is more popular and widely adopted by coconut farmers. He indicates that channel intermediaries play a major role in both assembling and equalization functions in marketing of coconuts.

Subburaj, B. and Singh, R.K. (2003) conducted a study on Marketing Mix for Coconut Products—Consumers' Perception. This study concluded that majority of consumers do not have high perception on coconut products. High income category finds difficulties with distribution mix, whereas other than the high income category finds problems with price mix. Promotion mix therefore is to be designed and executed in such a way that it could address the expectations and wants of target consumers in general and potential consumers in particular. Coconut products have high market potential for coconut products requires effective marketing strategies meant for both creating and fulfilling the demand.

Singh, R.K. and Subburaj, B. (2003) conducted a study on Highways—The potential markets for tender nut sale in Tamil Nadu to understand the nature and type of demand prevailing in the region stretching between Kanniyakumari and Bangalore during summer and non-summer days. The study concludes that majority of the motels that are located on the National Highways in Tamil Nadu sell tender nuts throughout the year. However, such motels are found larger in numbers in southern region than that in middle and northern regions of Tamil Nadu. Existence of pilgrimage as well as tourist places is attributed for the existence of the motels in large numbers that sell tender coconuts. On an average 193 tender nuts are sold per day. High sales occur during summer days, and it is moderate during spring days.

Anitha Kumari, P. and Jissy Geroge (2003) conducted a study on Gender Perspectives In Coconut Product Diversification—An Analysis among 30 Self Help Groups for women regarding their awareness and knowledge on coconut product diversification, utilization of women groups in starting the enterprises and the prospects they perceive in adopting them. They conclude that enterprises on coconut product diversification are women friendly and suited for group activities for additional income generation.

H.P. Singh (2003) conducted a study on, "Augmentation of Coconut Marketing" in Kerala. He found that the coconut industry is growing in terms of production. However, its share in oil and fats trade has consistently declined in the past four decades. Vast growth opportunities remain for the coconut industry, but the marketing strategy needs a reorientation to suit to emerging trend. Awakening and alertness on diversification of coconut, with a motive to recapture the market have to be provided priority. The best option with vast growth opportunities for product diversification and value addition exists.

Thomas Mathew, Hameed Khan and Shivapuje (2003) conducted a study on Coconut in Konkan with the objective of knowing area and production of coconut, trend in area, production, productivity of coconut in Konkan region comprising the districts of Thane, Raighad, Sindhudurg and Ratnagiri of Maharastra. This study reveals that among the four Konkan districts,

Sindhudurg ranks first in the area under and production of coconut accounting for 58.44 percent in are and 42.24 percent in production of coconut and the trend in area under coconut in Konkan region was steady, till 1995-96 and then there was an increasing trend with a sudden increase in 1995-96. Thane where the productivity of coconut was increased to 15,771 nuts per ha in 2001-2002 from the level of 5320 nuts in 1985-86, which is 196.44 percent more than the base year. This study concludes that coconut is primarily a small holders plantation crop in Konkan region.

Thomas Mathew, M. (2003) in his paper on Coconut products in support of health and the importance of coconut in Maharashtra felt that coconut is a food crop to a major segment Indians and plays a dominant role in their social and religious rituals. It is an important cash crop for more than 10 million farm families in the country besides providing job securities to equal number of agricultural labourers. It is one of the important fiber yielding crops to more than 15.000 coir and coir based industries in the country which provide employment opportunities to about 5.86 lakh labourers of which 80 percent constitute women. The industry contributes to an amount of Rs.395 crore as the export earning from the coconut products and by-products mainly coir and coir products. The crop also contributes 6 percent of the total edible oil pool of the country. Coconut production in the country has made great strides during the past two decades. Currently the crop covers an area of 1.89 million ha with an estimated production of 12.8 billion nuts per annum, which accounts for about 23.36 percent of the world production.

Haridass, R. and Muthuraj, M. (2004) conducted a study on Economics of coconut cultivation in Tamil Nadu—A comparative analysis of Tall and Dwarf Variety with the objective of to analyse and compare the cost and return structure of tall and dwarf variety of coconut. The results of the net present value, benefit cost ratio, internal rate of return and pay back period showed that dwarf variety of coconut is quickly yielding and economically viable than tall variety.

Subburaj, B. (2004) conducted a Research Study on "Markets of non-conventional coconut products" with an objective of studying demand, supply and consumers behaviour in Tamil Nadu. The study estimates that the marketed surplus at farm level is 1562 million nuts while it is 1111.35 million nuts at market level. Nearly 414.33 million nuts are consumed with in Tamil Nadu for culinary and religious purposes. About 969.63 million nuts are consumed by way of coconut oil, 50.08 million nuts as tender nuts and 56 million nuts for manufacture of non-traditional coconut products. The market surplus for the year 2003-04 is estimated at 35.64 million nuts.

Subburaj B. (2004) conducted a research study on coconuts in Regulated Markets aiming to assess the role Regulated Markets in marketing of coconut and coconut products in the state. The study reveals that the user and non-user farmers have high perception of the contributions of regulated markets in regulating and systematizing the agricultural marketing system.

Sum up

Although several attempts have already been made to gauge trend in coconut production and coconut marketing, studies dovetailing the role of market intermediaries coupled with production and marketing of coconut found hardly from the available literature. Hence the present study has been undertaken to fill this gap.

REFERENCES

1. Geoffrey Bastin, "Coconut Marketing: Issues for a Threatened Industry", *Cord,* July 1986, p. 20.
2. Raveendran, P., "Demand, Consumption Pattern and Consumer Acceptability of Desiccated Coconut", Processing and Marketing of Desiccated Coconut, Coconut Development Board, Cochin, April 1987 p. 25.
3. __________, "Marketing of Coconut Products in Andaman and Nicobar Islands", *Indian Coconut Journal,* December 1989, p. 14.
4. Ratha Krishnan, L., "Price Spread in Coconut Marketing", *Indian Coconut Journal,* March 1989, p. 20.
5. Hameed Khan, H. and others, "Improving The Coconut Production", *Indian Coconut Journal,* April 1990, p.2.
6. Punchihewa, P.G., "Coconut Industry—Current Situation and Prospects", *Indian Coconut Journal,* November 1990, p. 2.
7. Anitha Sharma, "Villagers can do well with better marketing", *Indian Farmer Times,* December 1990, p. 9.
8. George, M.V. and others, "Trend in Area, Production and Productivity of Coconut in India", *Indian Coconut Journal,* November 1991, p. 45.
9. Thomas Mathew, "A Study on the Consumption of Tender Coconuts in Tamil Nadu", *Indian Coconut Journal,* September 1990, p. 5.

10. Jaganathan, N., "An Economic Analysis of Coconut Farming in Anamalai Block of Coimbatore District, Tamil Nadu", 1991, p. 10.
11. Ron Harris, "The Coconut Industry Utilization and Marketing", *The Planter,* April 1991, p. 161.
12. Jos, C.A., "Cooperativization of Marketing and Processing of Coconuts—Problems and Prospects", Processing and Marketing of Coconuts in India., Cochin: Coconut Development Board, 1992, p. 93.
13. Profulla Das., "Coconut Marketing Problems, Prospects and Challenges in Kerala", Processing and Marketing of Coconuts in India, Cochin: Coconut Development Board, 1992, p. 134.
14. Ramakrishna Pillai.K., "Role of Cooperatives in Processing and Marketing of Coconuts in Kerala with special reference to MARKETFED", Processing and Marketing of Coconuts in India., Cochin: Coconut Development Board, 1992, p. 90.
15. Markose. V.T and Sree Kumar., "Coconut Based Rural Industries—An Overall Perspective", *Indian Coconut Journal,* December 1993, p. 2.
16. ——, "Coconut Development in the Last Decade", *Indian Coconut Journal,* November 1994, p.12.
17. Chengappa.P.G and others., "Methods of Disposal of Coconut and Economics of Processing Coconut at Farm Level", *Indian Coconut Journal,* April 1993, p. 5.
18. Marapandiyan, P., "Role of KERAFED in Coconut Processing and Marketing", Processing and Marketing of Coconuts in India, 1995, p. 92.
19. Narasimhappa., "Processing and Marketing of Coconut and its Products", Processing and Marketing of Coconuts in India, Cochin: Coconut Development Board, 1995, p. 129.
20. Haridas, R. and Chandran, C., "Marketing systems, costs, margin, price spread and marketing problems of coconut— A case study of coconut growers and traders in Tamil Nadu", *Indian Coconut Journal,* May 1996, p. 13.
21. Aravindakshna, M., "Challenges of the Coconut Industry in India and Strategy for Making it Competitive", Paper presented in the COCOTECH Meeting held at Kochi, July 1995, p. 1.
22. Thampan, P.K., "Coconut Diversification, a Must", The Survey of Indian Agriculture, Chennai : *The Hindu,* 1996, p. 99.
23. __________ and Venkitachalam,V., "Global Coconut Situation and Strategy for Development", *Coconut for Prosperity,* Cochin : 1996, p. 109.

24. Thomas, P.T. and Venkitachalam, V., "Investment Opportunities in Coconut in Tamil Nadu", *Indian Coconut Journal,* June 1997, p.4.
25. Venkattakumar, R., "A study on Socio-Economic Analysis of Commercial Coconut Growers" Unpublished Thesis, Tamil Nadu Agricultural University, Coimbatore 1997, p.47.
26. Subburaj, B., "Marketing of Coconuts", Research Monograph, Gandhigram Rural Institute, January 2000, p. 67.
27. P. Rethinam and others, "a study on problems and prospects of coconut industry in Andaman Nicobar Islands", *Indian Coconut Journal,* June 2001, p. 2.
28. M.K. Nair and M.K. Rajesh, "a study on coconut production and productivity", *Indian Coconut Journal,* June 2001, p. 13.
29. P. Rethinam and P.K Thampan, "a study on Reviving coconut based economy of Kerala state", *Indian Coconut Journal,* December 2001,p.2
30. NABARD, "Coconut Development in Tamil Nadu - An Ex-Post Evaluation Study in Coimbatore District", Evaluation Study Series No:11, NABARD, February 2002, p.9.
31. Thomas Mathew. M., "a study on Trade in Tender coconut - A future vista of Indian coconut industry", *Indian Coconut Journal,* July 2002, p. 1.
32. R.K. Singh and B. Subburaj, "a study on Pricing the coconut in Tamil Nadu - An Analysis", Indian Coconut Journal, December 2002, p.1.
33. B. Subburaj and R.K. Singh., "a study on Marketing Mix for Coconut Products - Consumers' Perception with the help of empirical study", *Indian Coconut Journal,* March 2003, p.6
34. R.K. Singh and B. Subburaj., "a study on Highways- The potential markets for tender nut sale in Tamil Nadu", *Indian Coconut Journal,* August 2003. p.6.
35. Anitha Kumari, P. and Jissy Geroge, "A study on Gender Perspectives In Coconut Product Diversification—An Analysis", *Indian Coconut Journal,* October 2003, p. 17.
36. H.P. Singh, "Augmentation of Coconut Marketing", *Indian Coconut Journal,* October 2003, p. 3.
37. Thomas Mathew, Hameed Khan and Shivapuje, "a study on Coconut in Konkan", *Indian Coconut Journal,* November 2003, p. 9.
38. Thomas Mathew, M., "Coconut Products in Support of Health and the importance of coconut in Maharashtra", *Indian Coconut Journal,* November 2003, p. 25.

39. R. Haridass and M. Muthuraj, "A study on Economics of Coconut Cultivation in Tamil Nadu—A comparative Analysis of Tall and Dwarf Variety", *Indian Coconut Journal*, May 2004. p. 3.
40. Subburaj, B., "Markets of Non-Conventional Coconut Products", Research Monograph, Gandhigram Rural Institute, January 2004, p. 34.
41. ——, "Research study on Coconuts in Regulated Markets", Research Monograph, Gandhigram Rural Institute, January 2004, p. 37.

3

Coconut Production—A Bird's Eye View

The coconut palm is considered as the tree of life. The palm is eulogized as kalpavriksha, the all-giving tree or the tree of Heaven and its fruit as *Lakshmiphal,* the fruit of wealth. Millions of families earn their livelihood from these trees by making use of its products and by products. The crop is gaining significance as a fibre crop, beverage and a medicinal crop. It is a traditional plantation crop grown in India for the last 3000 years and thus possesses the longest recorded history in the country (Subburaj, B. 2000). Its cultivation in the past was mainly confined to the coastal belts of west coast and east coast. From the coastal belts it has spread to interior areas and several states in the north and north east. The crop assumes considerable significance in the national economy in view of the rural employment and income generation. It is an environmentally benign crop offering favourable microclimate, nutrient conservation and recycling, adaptability to problem soils, control of natural weeds and sources of renewable energy.

Brief Botanical Description of Coconut

Family: palmae

Name: *Cocos nucifera* L.

It is a tall, stately, unarmed palm growing to a height of 12 m to 24 m. The trunk is stout, flexuous, annulate, rarely stands vertically, but makes a gradual curve, rising from a swollen base surrounded by a mass of adventitious roots. The stem is marked by rings of leaf-scars which are often not prominent at the base. Leaves are large, long, pinnatisect, borne of the crown; leaflets equidistant, 60cm to 90cm long, narrow, tapering, linear, lanceolate, coriaceous, flaccid; petioles stout, 90 cm to 150 cm long.

The Leaf

* Every tree has a crown of leaves oriented at the top of the trunk consisting of opened leaves and those in the bud in various stages of development.
* The number of leaves in the crown varies depending on the conditions prevailing, viz., method of cultivation, nutritional status of the soil, sunlight, etc.
* In adult trees, the crown comprises about 30 to 40 opened leaves.
* There are generally four sets of leaves in the crown.
* The first set consists of ten to twelve leaves counted from the oldest leaf, from the axils of which the bunches have already been harvested.
* The majority of these are of very little value to the tree having past their prime.
* The next set consisting of ten to fourteen leaves will be those supporting fruit bunches in the different stages of development.
* The third set includes ten to twelve opened leaves in the axils of which are found spadices in various stages of development.
* The soft and brittle bud in the centre of the crown is the most rapidly growing part of the tree.
* The bud is rich in phosphates and potash and has a 'nutty' flavour.
* It is considered a great delicacy.
* The number of leaves inside the 'cabbage' has to be generally ranging from the same number to one and a half times the number of leaves opened outside.
* The bud is protected by stipules, which form a sort of fibrous bandage binding the tender leaves.
* This fibrous sheath persists for some time till the leaves mature and become strong enough to withstand severe winds.
* The bud produces leaves in succession.

* The young leaf first appears in the centre of the crown as a pointed arrow and in this stage all the leaflets are held together.
* The top leaflets open out as the leaf grows and it takes four to five months to completely emerge out of the sheath.
* The leaf is considered to be fully open when the lowermost of the leaflets has emerged.
* When the tip of the leaf appears in the centre of the crown it is in line with the stem, i.e., almost perpendicular to the ground.
* It is then gradually pushed to one side to provide space for the younger set of leaves.
* As the leaf gets older, it bends down and by the time it dries up and is about to shed, i.e., within about two and a half to three years from the emergence of the leaf.
* The tip which once faced the sky now points downwards to the ground.
* The leaves are so arranged on the crown that each leaf gets the maximum amount of light.
* This is secured by an arrangement of leaves on the crown in five spirals which run in clock-wise or anti-clockwise direction giving rise to right of left spirals respectively.
* The phyllotaxy of each spiral is nearly two-fifths. Starting from the first leaf and making nearly two complete rounds of the stem the sixth leaf is reached.
* If the bunches hang towards the right of the petiole, the spiral is towards the left and when the bunches hang towards the left of the petiole, the spiral is towards the right.
* Even though the direction of the spiral varies in different trees, it remains the same throughout the life of any particular tree.
* The leaf consists of a rachis popularly known as leaf-stalk (or petiole) and the leaflets.
* The leaf-stalk continues as a mid-rib or rachis till the tip of the leaf where it merges into the green leaflets.
* The upper half of the leaf can withstand strains from above as well as from the sides.
* The leaf stalk attaches the leaf to the stem by means of a sheath in the form of a bracket firmly clasping the stem with its wings nearly round it, providing mechanical

strength to it to withstand considerable strain from the wind and the weight of the bunch.

* The centre of attachment is, however, not so strong, for if the joint at the wings is cut, it is quite easy to tear away the leaf from the stem which is still quite tender where it is protected by the thick butt of the leaf bracket.
* When young, the stipules from a part of the leaf stalk and surround the whole stem.
* In young trees these stipules frequently persist unbroken till the leaf dies.
* The shape of the leaf stalk is of importance and there is a certain natural variation in this.
* The upper surface is nearly flat or just sufficiently grooved to prevent water from running to its side.
* The groove down the centre of the mid-rib is of interest as this canalises sufficient water which is allowed to run down into the crown and from there down the stem to the base of the bole.
* The lower surface is a round keel which thickens out considerably at the base thus forming a supporting bracket.
* This rounded keel should be deep, as a shallow keeled leaf stalk is unable to support the weight of the green bunch.
* At the same time the leaf stalk should be wide so as to lesson the chances of the bunch from slipping off the support.
* The leaf stalk should also be short, as a short leaf stalk denotes a short bunch.
* If the stalk of the bunch is short, the bunch will be nearer to the stem and this brings about less strain on the leaf bracket.
* The number of leaflets on a leaf varies from 200 to 250. Where the number is small the leaflets are usually narrow and are farther apart, which is an undesirable character.
* The leaflets near the base of the leaf as well as those near the apex are much shorter and narrower than those situated in the middle.
* The longest leaflets are in the lower one-third of the leaflet-bearing region.
* It is thus clear that there is great variation in the area of the leaf surface in different trees.

* The attachment of leaflets at different parts of the leaf varies greatly, but in all cases it is so arranged that it throws water off the mid-rib down the leaflets so that it drips down at the ends.
* In the case of young leaves, these are more or less vertical and in this case any water falling on them runs down into the crown.
* The leaflet consists of a central strong mid-rib with a blade on each side.
* The blade is tough and leathery with green shining upper surface.
* The leaflets partially close longitudinally in the shade, cloudy weather or at night, while in full sunlight they remain open.

The Root

* The coconut palm has an adventitious root system as typical of a monocot, i.e., it produces numerous uniformly thick roots from the base of the stem almost throughout its life.
* The roots radiate from all the sides and are generally disposed to keep the direction in which they start.
* These are the main roots.
* The total number of such roots in a tree depends upon its age and the environment.
* On the poor sandy soils of the Malabar Coast (India), they may vary from 1,500 to 2,500 in a tree aged about 25 years and 3,620 roots in a 60 to 70-years-old tree.
* The number of roots counted on a bole usually varies from 4,000 to 7,000.
* Although numbers much higher than these are rare, it is interesting to note that as many as 11,360 roots have been counted on a bole of a very old tree at the Central Coconut Research Station, Kayangulam (loose sandy soil).
* The main roots measure 6 m in length on an average and in rare instances they may grow even up to 25 m. they are usually of uniform thickness of about 8 mm in diameter.
* Under certain environments (when mechanically or otherwise injured) they produce one or more branches which are almost as thick as the original roots.

* The main roots and the branch-roots are generally long-lived, while the rootlets, particularly the territories and their branches, are short-lived.
* However, the latter are replaced often during favorable conditions.
* Coconut roots are also provided with numerous lenticel-like outgrowths or ventilators which may be called 'breathing organs', 'breathing roots' or 'pneumatophores', which help in the exchange of gases.
* These white, pointed breathing roots are seen mostly on the main roots and sparsely on the rootlets.
* Normally, roots are not formed very much above the bole region even when buried in the ground.
* But, when the tree is old or when raised on water-logged areas, the roots commence to die from the apex upwards.
* When this stage is reached, the tree, in order to maintain itself, commences to form new roots above the normal root-producing zone at the base of the stem.
* These are termed aerial roots.

The Stem

* When the bole reaches the full stage of its development, the stem becomes visible.
* The thickness of the stem generally depends upon the vigor of the tree, hereditary variation and soil conditions.
* In the initial years of the growth of the tree the stem gradually increases in thickness and this thickness is maintained throughout.
* Growth of the stem may however, become thinner as the tree gets older are meets with unfavorable conditions.
* The unevenness thus produced, though slight, persists and this is a visual record of the vicissitudes through which the palm has passed.
* The stem of the coconut grows erect under normal conditions.
* But trees on the margins of the plantations often lean away towards light.
* In places where winds often prevail, the trees will lean along the directions of the wind.
* Any injury caused to the stem remains as the tissues of the stem have no capacity to repair the injury because

of the absence of meristematic activity.

* It can however withstand considerable amount of damage due to the presence of fibrous tissue in the trunk.
* These strands are generally much closer together towards the middle of the tree enabling it to withstand considerable lateral strain particularly in times of severe winds and gales.
* The coconut tree has only one growing point, the bud. If this bud is killed, the tree dies.
* The bud produces a succession of leaves.
* When the leaves grow old they are shed , leaving a rough pitted scar on the trunk which encircles the whole stem.
* From the counts of these scars the age of the trees can be approximately determined, as roughly 12 to 14 such successive scars left on the tree correspond to a year of growth of the tree.
* It is also possible to gauge the vigour of the tree from these scars.
* In a bearing tree the scars left by the inflorescence stalk may be seen on the stem in the middle above the widest portion of the leaf scar.
* The interspace between successive leaf scars may serve as an index of the bearing capacity of the tree or the environment under which it is grown.

Floral Biology

The Inflorescence

Inflorescences of the coconut are formed in the axils of every leaf of a bearing tree. Early workers on the coconut believed that inflorescences were formed in the axils of very third leaf. It is true that some axils fail to throw out inflorescences because the latter become abortive. But the production of such inflorescences has not been observed to have any definite relation with others. The coconut is a monoecious plant producing male and female flowers separately on the same tree. There are also palms which produce either completely male or female flowers. The tendency to produce completely male inflorescences is frequently observed when the palm produces the spadix for the first time in its life. Sometimes, this tendency is also manifested by trees which are given rest after a prolonged period of tapping for toddy. There is also considerable variation in the frequency of bearing such inflorescences. Some trees will produce their second or third inflorescence with female flowers in addition to male flowers, while others will not bear female flowers even in the sixth inflorescence. The coconut produces inflorescences in continuous succession. If one succeeds the other rapidly, or if the flowering period of each is sufficiently prolonged, it happens that before an inflorescence completes its flowering phase, the next one might be in flowering, bringing in chances of overlapping and consequent inter-spadix pollination. This frequently happens in dwarf palms and sometimes during the months of May, September and October in tall palms. The inflorescence appears at first in the axil of a leaf as a pear shaped flat structure. At this time, it is completely protected by two sheaths and is collectively known as the spadix. The outer sheath is thick and fibrous while the inner one is more yellow and often fibrous and somewhat flat in nature. In course of time, when the spadix is full grown, it becomes more cylindrical especially in the upper half and the distension of the inflorescence within causes great pressure on the walls of the spathe. The inflorescence consists of many flower bearing ramifications or spikeltes situated on a central axis or a peduncle. Its size varies from 0.75 m to 2 m in length depending upon the individual palm. When the inflorescence is enclosed within the spathe, the branches lie close to the main axis and the whole is so tightly packed that it is quite impossible for any of the individual flowers to open before the spathe splits. Sometimes, a few flowers open as soon as the spathe splits. As other times, none of the flowers open until the branches of the inflorescence have spread out. Each branch is fringed and crowded with numerous male flowers from the tip downwards and bears lower down one or more female flower, all the flowers being sessile or sub-sessile.

The Female Flower

Unlike the male flowers, the female flowers are comparatively few in number in an inflorescence. They may vary from 0 to 300 in each spadix depending on conditions prevailing, viz., nature of the tree, cultivation, manuring, season of the year, age of bearing, etc. At the time of opening of the spathe, the female flower is a small spherical body, about 1.3 cm in diameter with great resemblance to a small nut and is popularly known as a button The female flowers, like males, consist of six floral leaves which are thicker, imbricately arranged and tightly folded over the inner parts of the flower completely enveloping the pistil. From the tip of the pistil, there extend downwards three ridges which make the whole ovary look globosely three-sided, each side being provided at its tip with a groove. These three grooves meet in the centre of the apex and form the stigma. After the inflorescence opens, and before fertilization takes place, the mass of fibrous tissue continues to increase in size and ultimately forces apart the floral leaves disclosing only its rounded upper surface which is surmounted by a white nipple. This nipple is marked by three equidistant grooves which meet at its apex and thus divide it into three triangular sections. When the female flower is ripe, these three segments separate and stand erect as three teeth, exposing the stigmatic surface on which the pollen is to fall in order that the flower may be fertilized. These three teeth constitute the stigma. Normally it is impossible for fertilization to occur before the stigma is ripe and this does not happen until long after the opening of the inflorescence. When the fertilization is complete, the stigma ultimately turns brown and the tissues around it collapse, forming a small black, more or less circular area containing three shrivelled teeth at the apex of the young fruit. The six floral leaves increase slightly in size, and form the whorl of perianth leaves which persist at the base of the fruit till it matures and dries.

The Number of Female Flowers

The number of female flowers in the inflorescence is variable. There is a difference in the production of female flowers among the trees as well as among the inflorescence of the same tree. The variation among the trees of the same age and receiving the same treatment is genetical. The average number of female flowers per tree in the group "good" is about 151, for the group "medium" about 120 and for the group "poor" only 41.

Seasonal variation in the production of female flowers

There is much variation in the number of female flowers produced during the different months. This difference is partly due to the differences in the number of spadices opened in the different months. But when the average number of female flowers per spadix are compared, similar differences are also noticeable. It is generally high during the months of March to May, the highest being in May.

The Male Flower

The male flowers exceed the female ones in the same sapdix and may vary from a few hundreds to thousands depending on the number of ramifications in the spadix and the length of the flower bearing region. They are the first to open. Each male flower has six yellow perianth or floral leaves arranged in two whorls, the inner three alternating with the outer. The inner perianth parts are more than three times the size of the outer ones. Enclosed in this floral envelope, three are six hammer shaped stamens, which yield large quantities of powdery yellow pollen. In the centre of each flower, there is a rudimentary or abortive pistil which divides at its apex into three teeth, each bearing a nectar gland, the nectar of which attracts ants, bees and other insects. Rarely this rudimentary pistil is absent. The flowers on the upper spikes of the inflorescence and those situated on the sides of the female flowers are the first to open and commence opening from the apex of the spike and extends downwards. Sometimes few flowers may open out of order. The flowers bloom throughout the day, most of them blooming from 8 A.M to 10 A.M.

The Pollen

When the anthers are fully mature, the pollen sacs burst along two longitudinal slits which coincide with the partitions of the pollen sacs and shed their pollen before the opening of the male flower. The pollen grains are spherical and smooth when very fresh. On exposure for a few seconds, they turn ellipsoidal with a longitudinal groove in the middle and they measure about 0.063 mm in length and 0.020 mm in breadth at the centre. In the mature pollen grain, there are three nuclei one of which is larger than the remaining two.

The Fruit

The fruit of the coconut, botanically known as a fibrous drupe and popularly the 'nut', consists internally of the endospermous kernel with the embryo embedded in it and externally protected by the fruit coat known as the pericarp which consists of three distinct and well defined regions, viz., the exocarp or epicarp, the mesocarp and the endocarp. The outermost region of the fruit coat is the exocarp (or epicarp) which consists of a tough smooth and hard fibrous skin where as in the ripe fruit it assumes green, red, yellow or brown colour. In the young fruit, particularly until the fruit is well devleoped, it varies in colour depending on the general colour of the parent tree. It may assume various shades, viz., bright red, pale yellow, pale green, red brown, brown green or dark green. Immediately below this is the fleshy portion known as the mesocarp which in the young fruit is astringent and in rare instances sweet and edible. When the nut develops and matures, this region turns fibrous. The thickness of this region is variable ranging from 2 cm to 15 cm. Below the fibrous covering is the endocarp which when the fruit matures, develops into the shell. This hard shell in the cross-section of the fruit is more or less triangular in shape and has three ridges on the outside which alternate with the ridges on the fruit. At the basal end of the fruit where it is attached to the fruit stalk, there are three eyes, one on each carpel, normally two of which are quite hard and the third one soft. One side of the triangular cross-section or the carpel of the fruit is narrower than the other two sides. On an examination of the nut, it is found that one of the three segments has a wider angle than the other two. It is in this widest segment that the soft or germinating eye is situated which is easily recognised on the nut as it is bigger than the other two. The fibre over this eye is also softer and generally less compact than elsewhere. The seed or the kernel or meat of the fruit is well protected by the hard shell and lies immediately below it. Placed between the meat and the shell, is the seed coat which is firmly attached to the meat. The meat in the ordinary nut is about 13mm thick, but occasionally the thickness may go up to 20mm. In dwarf and ornamental varieties on the other hand, the kernel is generally very much thinner. At the apex of the nut, immediately below the soft eye and embedded in the kernel lies a peg-like body, the germ or embryo, which, on germination, commences to grow and forces in way through the soft eye. Within the kernel is a cavity which in the unripe fruit is completely filled with the liquid popularly known as 'coconut water'. As the fruit ripens, this water gets very much reduced filling the cavity only partially. When stored after harvest, the water is completely absorbed in about five months.

Varieties of Coconuts

Although good varieties of coconuts have been known to exist in the different coconut growing countries of the world, no serious attempts were made to collect them and study them in detail at a representative centre with a view to classify them systematically. Most of the varieties are generally known by the name of the locality where they are grown. In some cases, the same variety is known by different names in different countries. About 50 years ago some of the important varieties from reputed coconut tracts of India and nearby countries were obtained by the Departments of Agriculture of Madras and former Travancore - Cochin States and tried at their respective coconut farms in the

West Coast of India (Kerala). Variety is generally a term designated to denote a single strain or a group of strains which distinctly differ in structural and functional characters from one another or a group of the same species which can be depended upon to reproduce itself true to type. The cross-pollination prevailing in the coconut has given rise to a highly variable progeny of palms. The promiscuous crossing occurring in the coconut is responsible for giving rise to a large scale multiplication of single genetically heterozygous varieties widely varying from each other or groups of closely related varieties. In certain regions particularly in some of the islands isolated from each other certain type suited to the locality with well defined, relatively constant phenotypic characters were found evolved possibly through generations of natural selection. Coconut being a seed propagated crop, requiring an unusually long period of pre-bearing life, the evolution of promising varieties through breeding became a difficult and time-consuming project

Dwarf Palm

As the name indicates, the dwarf coconut is small in stature and commences bearing earlier than the tall variety. Dwarf coconut palms flower as early as the third year after planting and come to regular bearing in the ninth year, i.e. in about six years of first bearing. The origin of the dwarf variety has not been determined so far with any degree of accuracy. These are mutant from the tall or common form due to some change in genetic factors taking place at wright intervals leading to production of palms with pronounced size differences. These were reported to occur where large coconut areas exist and in widely distributed places. Although dwarf palms came into bearing early and yielded well, the copra was inclined to be softner, more pliable and leathery and not quite of such good quality as in that from tall From these it is difficult to prepare hard copra and usually a large proportion of wrinkled, distorted and rubbery copra is produced.

The Dwarf Variety

The Dwarf or short variety which has three distinct forms, namely, those producing green, orange and yellow nuts, is liked by some planters for its earliness in bearing and short stature and attractive colour of the nuts. The Dwarf palms are commonly known as Nicobar or Andaman Dwarf and also by certain local names. They begin to yield nuts in about three to three and half years after planting and throw out bunches with fairly large number of nuts. The palms are, however, irregular in bearing and susceptible to drought. The bunches with attractive, coloured nuts are generally used for decorative purposes. The nuts are small in size and ovoid or round in shape. They are often harvested in the tender nut stage for the sweet water which provides a cool and refreshing drink. The copra obtained from the nut weighs about 3 oz (85 gm) with 65 per cent oil content. It is leathery and is not of much commercial importance. For these reasons it is considered as an uneconomic variety. Dwarf types from Malaya, Nigeria, Ceylon and certain other places have been introduced at the Central Coconut Research Station, Kasaragod and are under study.

Tall Palm

This is the ordinary or the common tall variety of palms most extensively grown on a plantation scale in all coconut tracts of the world. It is a long lived palm living generally to an age of about 80 to 90 years, although sometimes older trees are also met with. The palms thrive well under different soil conditions varying from littoral sands to red loams and laterites and also under varying climatic and rainfall conditions. It is observed to grow well up to an altitude of 3,000 ft. above the sea level. It is fairly resistant to diseases and pests. The tree attains a height of about 15cm to 18cm or more. It begins to bear in about 8 to 10 years after planting. The palms of this variety are generally cross-pollinated in nature, although, in summer months there exist chances of self pollination due to the overlapping of spadices. After pollination, the nuts mature in a period of 12 months. The nut is medium to big in size varying in shape from spheroid to linear-oblong and with colours varying from green, yellow and orange to shades of brown. The quality and quantity of copra form nuts of this variety are satisfactory. About 6,000 nuts yield a ton of copra. These are generally the palms recommended for large scale planting as they are superior to the dwarf palms

The West Coast Variety

Of the varieties of coconut available in India, the West Coast variety, which is otherwise known as the ordinary or common Tall variety, is the one that is extensively cultivated in all the important coconut tracts of India and is of commercial importance. This variety is found to grow well in littoral sand as well as in the interior and up to an altitude of about 3,000 feet above sea level. It has been in cultivation in India from very ancient times and may, therefore, be considered as indigenous to the country. The West Coast variety is a long-lived, hardy, multipurpose palm, yielding nuts, copra, oil and fibre of good quality. The tree also yields, on tapping, good quantity and quality of coconut juice or toddy which can be fermented or converted into jaggery or sugar. In this variety, growers recognize different sub-varieties or forms based mostly on the colour and shape of nuts or bearing capacity. The nuts of this variety are generally of medium size, varying in shape from spheroid to linear with colour varying from green, yellow, yellow-orange to shades of brown. Some of the forms show variation in the thickness of the husk and thickness of meat or kernel. Trees yielding large number of medium-sized round nuts in almost every leaf axil without any tendency for the bunches to buckle or droop, are considered the best for planting on a large scale. The West Coast variety commences to yield in about six to eight years after planting when grown under favourable conditions. The time of first bearing may, however, be prolonged to ten to fifteen years or even more if conditions are unfavourable or adverse. The yield varies according to the ecotypes selected for cultivation and the conditions under which they are grown. The average yield of copra per nut is about 142 gm and the oil content in the copra is about 72 per cent by chemical extraction on dry basis.

Dwarf Varieties in India

Chowghat dwarf

The dwarf types, Chowghat green and orange are delicate and shortlived. They are easily susceptible to diseases and pests. The out-turn of copra is low and the quality is also poor. Chowghat dwarf green maintains high purity, while in dwarf orange both self and cross pollination occur.

Laccadive and Maldive dwarf

These are dwarf types of the Laccadive and Maldive Islands. They yield on an average not more than 50 nuts per tree per annum. They give good copra both in quality and in quantity. The copra content per nut is about 115 gm to 145gm and the percentage of oil in the copra is about 71.

Andaman dwarf

This is a dwarf type from the Andaman Islands. It is a poor yielder. The copra content per nut is about 145 gm and the oil content in the copra is about 63 per cent.

Gangabondam

This belongs to the 'medium dwarf' or 'semi-tall' type. The time of first bearing occurs in about four years after planting. The nut matures in eleven months. It is a prolific yielder. The nuts are elongated in shape, medium sized and have fairly thick kernel. They give good quality copra. The copra content per nut is about 230 gm and oil percentage in the copra is about 72.

Tall Varieties in India

Kappadam

This is allied to the ordinary tall type, but is more robust in all characters particularly in the size of the nut which is one of the largest on record. The shape of the nut is broadly ellipsoid. The yield is rather low being about 60 nuts per palm per year. Quantity of copra is high and copra is thick and hard. This type is met with in parts of South Malabar (India).

Andaman Giant

This is a type from the Andaman Islands. The palms are of tall stature, massive proportions and majestic appearance. The nuts are green and large, perhaps one of the largest on record but the yield is poor. Though the content is high, the copra is thin and of poor quality. It comes to bearing in about eight years after planting. The palms are resistant to diseases.

Andaman

This is a medium type. It resembles the Andaman Giant in all respects, except in stature. It is a good yielder of toddy.

Laccadive ordinary

This type resembles the ordinary tall variety. The average yield under ordinary conditions is about 127 nuts per palm per year. The copra content is 155 gm and oil content in the copra is 72 per cent.

Laccadive medium

In this type the nut is comparatively smaller in size than Laccadive ordinary. The average yield is about 182 nuts per palm per year and the average copra content is about 85 gm to 115gm with an oil content of 75 per cent.

Laccadive micro

Here the nuts are very small compared to those of the above mentioned types. The yield is about 160 nuts per tree per year. The average copra content is about 80gm to 95gm per nut with an oil content of 75 per cent.

Thairu thengai

There are certain palms among the tall type of the West Coast, which are locally known as Thairu thengai (curd coconut). In the nut there is no milk but it is completely filled with a jelly of the consistency of thick curd from which the name is derived. The kernel is not hard to make copra but it is good for eating. These nuts do not germinate but a few in every bunch produced by such trees are quite normal and when planted may give rise to trees of this type.

Kaithathali

Among the tall palms of the West Coast three is a variety called Kaithathali which is rare. It has a soft fleshy husk. In the tender nut, the fibres are so poorly developed that the husk is sometimes eaten raw and is said to be a very good antidote against sea-sickness.

Spikeless

This variety is quite distinct from the ordinary coconut palm in having no branches or spikes in the inflorescence. In this variety, femaleness is most and maleness least expressed, because the number of male flowers is as low as 50. The setting percentage is very low. The mature nut is smaller than the ordinary variety. It is found that about 50 per cent of the progeny of this variety breed true to type. It is also found to cross freely with other coconut varieties. This variety is found to be useless for tapping. It has been described as a distinct botanical variety, i.e, variety spicata.

Gangabhavani

This is a distinct type from Andhra. The nuts are big and oblong in shape. This type gives good tender nuts with sweet water. The copra content per nut is about 168 gm.

Verri kobbari

This type from Andhra is found to be almost similar to the

Laccadive small. The nuts are small, oblong and tapering both ends. The copra content per nut varies from 85 gm to 115 gm.

Rangoon kobbari

This type from Andhra appears similar to the Andaman Giant. The nuts are big and oblong. The copra content per nut is very high varying from 255 gm to 280 gm.

Hybrids

Some progenies of the Dwarf Orange type which are considered as natural crosses are found to turn out to be Semi-Talls and give good yield of nuts and copra of good quality. The seedlings which give rise to these off-type palms are conspicuous in the nursery by their vigorous growth, height, early splitting of leaflets and petiole colour, which are quite distinct from those of the pure Dwarf seedlings. Artificial crosses between the ordinary Tall variety and the Dwarf variety have resulted in the production of economic hybrids. The Tall (female) X Dwarf (male) hybrids are produced at the Coconut Research Stations at Kasaragod and Nileshwar. These hybrids combine in them the early bearing character of the Dwarf with the desirable copra character of the Tall. They are also short in stature and give fairly high yields. The nuts, however, are small when compared to those of the Tall variety. Consequently, the hybrids may not be liked by the coconut growers who realise a good income from the sale of raw husks required for retting purposes for the production of good quality coir yarn. Of the two Dwarf types, Orange and Green, the Orange Dwarf appears to be a better male parent for the production of promising hybrids. Crosses between some of the Semi Talls and Talls have also given rise to economic hybrids. Studies made so far have indicated the need for selection of the parents on the basis of economic transmissible characters for the production of good hybrids. It has however, to be pointed out that the progenies of hybrid palms should not be indiscriminately planted as they will segregate for the parental characters and give rise to undesirable dwarf palms.

Godavari Ganga (Hybrid)

It is a hybrid between East coastal tall as female parent and Ganga bondam as male parent. This variety is developed by scientists of Agricultural Research station, Ambajipet, up to now 2 lack seedlings are supplied for cultivation in Andhra Pradesh. On National wide this variety can be recommended for A.P and

Tamil Nadu states. This variety starts bearing from 6th year and yields on an average of 140 fruits per tree. This hybrid contains copra of 15 gram and oil content of 68%.

Other Varieties

Laccadive Ordinary

This variety resembles the ordinary West Coast variety. The nuts are medium-sized. The yield per tree per annum is about 124 nuts. The copra content is about 142 gm and the oil content in the copra is 72 per cent.

Laccadive Small

The palm resembles the ordinary West Coast variety in stature, but the nuts are definitely small-sized and spheroid or linear in shape. Large number of nuts are produced in a bunch and the copra is of good quality. The palm is, however, an alternate bearer and produces not less than 150 nuts per tree per year. The copra content is only 57 gm but the oil content of the copra is as high as 75 per cent.

New Guinea

This is a robust palm with tall and stout trunk. The nuts are large, spheroid or ellipsoid in shape, the colour varying from green to brown. The nuts contain plenty of sweet water in the tender nut stage. The quality of the copra is not as good as that of the West Coast variety. The palm yields, on an average, in the gravelly soil of the West Coast, about 65 nuts per tree per year. It has a copra content of about 227 gm and an oil content of 66 per cent.

Cochin China

This is a robust palm giving nuts which are large-sized, spheroid in shape and coloured green to shades of brown. The water in the tender nut is sweet and plentiful. It yields about 86 nuts per tree per year. The copra is of fair quality and weighs about 227 gm per nut and has an oil content of 66 per cent.

Java

This is a tall variety with fairly stout trunk. The nuts are medium to large in size and round of somewhat elongated in shape. It yields about 95 nuts per tree per year. The copra is of good quality. It has a copra content of 198 gm and an oil content of 66 per cent.

Siam

This is a fairly robust palm. The nuts are green in colour and medium to large in size and ellipsoid in shape. The water in the tender nut is sweet. The palm yields about 50 nuts per tree per year. The copra is of good quality and weights about 227 gm per nut. The oil content of copra is 74 per cent.

Origin of Coconut Farming

The inferences from Timeline and Trend change exercises carried out among villagers indicate that coconut palms were found in villages even 135 years from now. One hundred-year old palms that can be seen even today, serve as evidence. They are indigenous tall varieties located mostly on the farmwell beds and on either side of the space provided for traditional method of lifting water from their wells for irrigation by using motes and bullocks. Big farmers were reported as the early adopters. They raised coconut palms amidst other benevolent trees in the flood-prone and semi-marshy lands. However, they did not consider coconut palms as a crop.

Some 40-50 years later, coconut palms had acquired currency in the village economy. There were changes in the cropping pattern. Coconut palms were considered a crop. All categories of farmers raised coconut palms on the periphery of their cultivable land as the peripheral palms neither affected the ground crops nor required extra attention from farmers. This resulted in increase in coconut palms. Farmers realized the economies in coconut cultivation as they got assured income.

Trends in Coconut Production

Historical transect on the significance of coconut palms, pattern and extent of coconut cultivation, production enhancement operations, productivity and such other aspects depicting trend in coconut cultivation are discussed.

Table 3.1
Trends in Coconut Production

S.No.	Period/Details		100 years before		50 years before
1.	Adopters	o	Big farmers	o	All categories of farmers
2.	Number of palms	o	50-100	o	70-250
3.	Species of palms	o	Indigenous tall variety	o	Indigenous tall variety
4.	Location of palms	o	Banks of the stream and wellbed	o	Cultivable land & farm periphery
5.	Type of cultivation	o	As a palm amidst horticulture groves	o	Mostly as peripheral crop
6.	Reasons for cultivation	o	As a crop in the flood prone and marshy land as shadow palm	o	As a source of income
7.	Production enhancement operations	o	Nil	o	Watering

(Coconut Development Board- Technology Mission, Kochi, 2004.)

Table 3.2
Trends in Coconut and Byproducts

Products	100 years before	50 years before
Coconuts	➢ Domestic and ceremonial use besides sold in the market	➢ Domestic consumption
Coconut oil	➢ Domestic use for edible and non edible purposes	➢ Domestic and commercial use for edible and non edible purposes
Tender coconuts	➢ Ceremonial use and consumption	➢ Domestic consumption
By-products	➢ Domestic use	➢ Domestic use
Development of coconut based agro industries	➢ Nil	➢ Development of fibre units

(Coconut Development Board- Technology Mission, Kochi, 2004.)

Coconut Palm in the World

The coconut palm is essentially a tropical crop and is grown in countries located in the tropical region between latitudes 22° N-27°S. The distribution of palm extends over most of the island, peninsular and coastal areas and even in some places outside the tropic zone. The main coconut growing areas in the world are located in Asia, Oceania, the West Indies, Central and South America and West and East Africa. In Asia, it is rarely found in the north of India. In India, the palm grows as far north as Lucknow. In Africa, the northern limits are Cape Verde (15° N) on the West Coast and Djibouti (11° 30″N) on the East, while the Southern limits are Mossamedas (15° S) on the West Coast and the Zambesi river (19° S) on the East. In the pacific, the palm is found as far north as the Bonin Islands (26° N) and as far South as the Pitcairn Islands (25° S). In South America the palm is found in Brazil at a ltitude of 27° S and in North America in Florida and Bahamas at a latitude of 25° N and in Mexico. It is grown in nearly ninety countries located in the tropical belt of the world (Satyabalan, 1993).

The coconut palm is the most extensively grown nut in more than 90 countries in the world. It is grown in an area of about 12 million hectares in the world, which account for about 55.569 billion nuts. More than 84 percent of the production is from Asia (APCC 2006). The Philippines, Indonesia and India account for more than 76 percent and 72 percent in the area under coconut and the production of nuts respectively. The coconut palm is essentially a crop of small farmers and is, perhaps, the only source of livelihood to millions of farmers in almost all the countries where it is grown (Thampan 1988). The coconut industry provides livelihood for one million, three million, three million and 10 million people living in Malaysia, Indonesia, Sri Lanka and India respectively (Punchihewa 1991). It supports more than 80 million small and marginal farmers in the world (Singh 2000). Indonesia, Philippines, India and Sri Lanka are the major players in the global coconut scenario, and in all these countries the coconut exerts profound influence in the economy. In India, coconut is not only associated with rural economy but is interwoven in cultural heritage.

Among coconut growing countries, Asia and the Pacific regions produce 48448 million nuts from 10.65 million ha sharing 87.19 percent in production and 90.26 percent in area. Africa and American regions produce 7121 million nuts from 1.15 million ha sharing 13.41 percent in production and 9.74 percent in area.

Table 3.3

Continent Wise Area and Production

Region	*Area*		*Production*	
	(M. ha.)	*% share*	*(Million nuts)*	*% share*
Asia and Pacific	10.65	90.26	48448	87.19
Africa	0.63	5.34	2232	4.62
America	0.52	4.40	4889	8.79
Total	11.8	100.00	55569	100.00

Source: APCC and FAO Production Year Books-2006.

During the decade, there has been 5.64 per cent increase in area, 17.41 per cent increase in production but only 11.15 per cent increase in productivity. (Table 3.4) However, production continued to increase till 1995 and then there was a sharp decline both in production and productivity up to 2000. Declining trend in productivity after 1995 could be attributed to increasing number of old and senile trees. But the year 2000 onwards there was a steep increase both in production and productivity due to the high yield varieties.

Table 3.4

World Area, Production and Productivity of Coconut for the last 16 years

Year	*Area (Mill.ha)*	*Production (million nuts)*	*Productivity (nuts per ha.)*
1990	11.16	47328	4240
1991	11.23	46592	4132
1992	11.19	47595	4254
1993	11.12	49000	4399
1994	11.34	51130	4505
1995	11.82	54021	4569
1996	12.19	52924	4343
1997	12.34	54603	4423
1998	12.53	53649	4281
1999	12.75	54767	4295
2000	12.72	54802	4308
2001	12.80	56852	4441
2002	12.12	58104	4794
2003	12.12	58426	4821
2004	12.12	57603	4753

2005	12.16	59569	4899
2006	11.79	55569	4713
% increase over 1990	5.64	17.41	11.15

Source: APCC and FAO Production Year Books-2006.

Indonesia ranks first in coconut production and area by contributing 14,984 million nuts from 3.81 million ha, sharing 32.31 per cent in area and 26.96 per cent in production. Coconut in Indonesia is grown mainly by small holders and forms a major economic activity of about 2.5 million farm families. In Philippines, coconut is grown in 3.24 million ha. producing 12,600 million nuts, ranking second in area and third in production. A coconut industry support nearly one-third of Philippines' population and it is basically a small holders crop. The percent share of Philippines in area and production is 27.49 m ha. and 22.68 m. nuts respectively. In India, the area and production of coconut are 1.93 million ha. and 12,600 million nuts respectively. The crop supports 11 million people, mainly living in the coastal tracts in the country. It contributes Rs.72, 000 million to the Gross Domestic Product and Rs.3,620 million to the foreign exchange earnings mainly from export of coir products.

In Sri Lanka, area under coconut comprises 20 per cent of agricultural land. It is the fourth largest producer contributing 2,784 million nuts from a total area of 0.39 million ha with a per cent share of 3.30 and 5.00 in area and production respectively. Coconut is the third largest foreign exchange earner of the country. It provides direct employment to 0.17 million people. The Vietnam although its percent share is only 1.23 percent, coconut is becoming very important in her economy.

The other countries, which contribute significantly to the world production of coconut, are Bangladesh, Papua New Guinea and Myanmar, smaller islands viz., Vanuatu, Solomon Islands, Samoa, Fiji, Palau and Micronesia. Although these countries have small share in coconut production, the economies of these countries are significantly benefited by coconut.

Indian Scenario

Throughout India coconut palm is eulogized as kalpavriksha or the 'tree of heaven' and its fruit as lakshmiphal, the fruit of wealth. Coconut is associated with heritage, culture and economy of the country. Coconut palm products especially the nuts are used in India on varied occasions. Even though uses of palm products such as leaves, inflorescence, toddy etc., are confined to traditional coconut growing areas, the nut is used throughout India. In the State of Kerala where over 50 per cent of the area under coconut is concentrated, there is hardly any ritual or ceremony in which coconut palm products are not used. There is a saying in Kerala "*Chathalum Thenga, Pettalum Thenga*" which means "birth or death, coconut is a must". The name Kerala *(Keralam)* is derived from the palm-*Keralam* means the land of coconut palm (Aravindakshan and Nandakumar, 1996).

Table 3.5
Area, Production and Productivity of Coconut in Major Coconut Growing Countries (2006)

S. No.	*Country*	*Area (Mill.ha.)*	*% share*	*Production (Million nuts)*	*% share*	*Productivity (Nuts/ha.)*
1	Indonesia	3.81	32.31	14984	26.96	3932
2	Philippines	3.24	27.49	12600	22.68	3888
3	India	1.93	16.38	12832	23.09	6648
4	Sri Lanka	0.39	3.30	2784	5.00	7138
5	Tanzania	0.31	2.64	465	0.83	1500
6	Brazil	0.28	2.37	3483	6.28	12439
7	Papua New Guinea	0.26	2.20	812	1.46	3123
8	Thailand	0.22	1.86	1248	2.25	5672
9	Vietnam	0.13	1.10	681	1.23	5238
10	Others	1.22	10.35	5680	10.22	4655
	Total	11.79	100	55569	100	

Source: APCC and FAO Production Year Books-2006.

Coconut is considered divine by the people in Tamil Nadu also. Coconut is called *'Thennam Pillai'* and treated like an young child. To destroy or discard the young palm is a sin (Chettiar, 1973). Coconut is considered equal to the sacred cow. It is believed that instead of giving a cow as a gift, a coconut would suffice. It is offered to gods as a token of devotion while celebrating *Pujas* (rituals to propitiate gods) in temples and houses. In Hindu religious rituals, coconut is also placed as a symbol of deity. Genesha or Vigneswara is the most popular God worshipped by the Hindus. For the *homam* (sacrifice) to propitiate Lord Ganapati, small pieces of coconut meat or copra are offered to sacrificial fire. Grated coconut and coconut milk are also used in the preparation of many puddings offered to deities in temples. In Punjab, when a child cuts his first tooth, fresh coconut is distributed (Singh, 1971). Varuna, the Lord of seas, is represented by a pot of water with coconut placed over it in religious rituals. Breaking dehusked coconuts by throwing on a granite stone is a common offering to Lord Ganapati and Lord Ayyappa in temples. Vishnu is an important festival for Keralites which heralds the beginning of the traditional New Year. Important ceremony connected with this festival is *Kanikanal* (to see auspicious things first after waking up in the early hours of the day) for which coconut is an integral part.

Tender coconut called Ilaneer in south India is also used in ceremonies and rituals. *Abhisheka* (pouring over deity) using tender coconut water is an important offering in temples dedicated to Lord Siva and Lord Subramania in the country. Drinking tender coconut water on *Thiruvathira* day (a festival of women) by the women is popular in Kerala. Snake worship was popular in Kerala and for ceremonies to propitiate snakes also tender coconut water was essential, coconuts and other coconut palm products were also used in the rituals (Nandakumar, 1985).

Hindu wedding in India varies in style and content according to the location and caste. But whatever be the location and the caste and whatever be the rites and ceremonies, the fruit of the coconut palm has an important place in all wedding ceremonies. In marriages of Brahmins a *Kumbham* (a coconut placed over a pot of water) is required. Coconuts are also given to guests who attend the marriage ceremonies. In Karnataka there is a practice of presenting Phalathamboola to guests attending marriages and during the visit to one's house during festivals. Phalathamboola consists of one coconut, an arecanut and betel leaves. In Manipuri marriage, bride's father offers bridegroom rice, banana, a coconut etc., in a plate. In Orissa, the bridegroom carries a coconut with him and keeps it till he offers it to the *Homakunda* (fire pit) set in his house on the fourth day of the marriage. In Parsi marriages, when bride as well as bridegroom are received, a coconut is waved over their heads and then it is broken in front of their feet. In Punjabi marriages when all the initial negotiations are completed, the girl's father sends a coconut, dry dates, sugar and some money as first auspicious gifts to the boy's family. (Edachal, 1963).

Spindle leaf *(Kuruthola)* of coconut palm is also considered very auspicious and used for decorations in temples, houses and public places during festive occasions. Spindle leaf of coconut palm is considered very auspicious also by Christians. Hosana festival or Palm Sunday is celebrated by the Catholics in India.

Other coconut palm products such as husk, inflorescence, spathes, oil, toddy and shell are also used in rituals and ceremonies.

Foods which are offered to gods are cooked in Kerala using coconut husks and spathes as fuel, since they provide slow fire and steady heat. For homam, to prepare sacrificial fire, husks and spathes are used. Coconut toddy is also considered sacred. It is offered to deities in some temples in Kerala and is used in rituals connected with sorcery.

Trend in Area, Production and Productivity of Coconuts in India

There is distinct difference in the pattern of distribution of the coconut in the country. Bulk of the coconut production is contributed by the four southern states viz., Kerala, Karnataka, Tamil Nadu and Andhra Pradesh. Other traditional coconuts growing States/Union Territories (UT) in the country are Orissa, Maharashtra, West Bengal, Goa, Lakshadweep and Andaman and Nicobar Islands. The crop now is grown in favorable locations in the Central, North and North-Eastern regions of the country. Of the present area of 1.90 million ha in the country, the contribution of Kerala in area is 0.81 million ha. followed by Karnataka (0.40 million .ha.), Tamil Nadu (0.38 million ha.) and Andhra Pradesh (0.1 million ha.) sharing 43.02 per cent,21.28 per cent, 20.14 per cent and 5.33 per cent respectively, which together accounts for 89.77 per cent of the total area under coconut in the country. In the production of coconut, Kerala tops the list with 5641 million nuts with a share of 38.26 per cent followed by Tamil Nadu with 4968.20 million nuts (33.70 per cent), Karnataka with 1635 million nuts (11.08 per cent) and Andhra Pradesh with 1119.26 million nuts (7.59 per cent).

Table 3.6

Coconut Production in India: State-wise (2007-2008)

States /UTs	*Area ('000 Hectares)*	*% Share*	*Production (Million nuts)*	*% Share*	*Productivity (Nuts/ha.)*
Andhra Pradesh	101.32	5.3	1119.26	7.59	11047
Assam	19.00	0.99	136.00	0.92	7158
Goa	25.50	1.34	127.60	0.86	5004
Gujarat	16.40	0.86	138.30	0.94	8433
Karnataka	405.00	21.28	1635.00	11.08	4037
Kerala	818.80	43.02	5641.00	38.26	6889
Maharashtra	21.00	1.10	175.10	1.19	8338
Nagaland	0.90	0.04	0.20	0.01	222
Orissa	51.00	2.67	275.80	1.87	5408

Tamil Nadu	383.37	20.14	4968.20	33.70	12959
Tripura	5.80	0.30	11.40	0.07	1966
West Bengal	28.60	1.50	355.50	2.41	12430
A and N Islands	21.60	1.13	80.60	0.55	3731
Lakshadweep	2.70	0.14	53.00	0.36	19630
Pondicherry	2.20	0.11	26.60	0.18	12091
All India	1903.19	100.00	14743.56	100.00	7747

Source : Directorate of Economics and Statistics, Ministry of Agriculture, Govt. of India, 2008.

The country has recorded a sizeable increase in area, production and productivity during the last one decade. The area increased from 1.51 million ha to 1.90 million ha. with an increase of 0.39 million ha and production from 9,700 million nuts to 14,743 million nuts with an increase of 5,043 million nuts. The productivity increased from 6,407 nuts to 7747 nuts per ha. Thus it could be seen that despite the undulating growth behavior exhibited in production by a few states, the Indian scenario presents a healthy trend in expansion in area and increase in production for the last few years.

Table 3.7

Trend in Area, Production and Productivity of Coconut in India

Year	*Area ('000 ha.)*	*Production (Million Nuts)*	*Productivity (Nuts/ha.)*
1990-1991	1513.90	9700.20	6407
1991-1992	1528.90	10079.60	6593
1992-1993	1537.70	11240.90	7310
1993-1994	1635.10	11974.70	7324
1994-1995	1713.80	13299.60	7760
1995-1996	1830.90	12952.30	7074
1996-1997	1890.80	13061.00	6908
1997-1998	1861.00	12717.30	6834
1998-1999	1754.50	12535.90	7145
1999-2000	1768.10	12129.00	6860
2000-2001	1823.91	12678.40	6951
2001-2002	1932.30	12962.50	6709
2002-2003	1921.80	12535.00	6523
2003-2004	1933.70	12178.20	6298
2004-2005	1935.00	12832.90	6632

2005-2006	1946.80	14811.10	7608
2006-2007	1936.80	15840.40	8179
2007-2008	1903.19	14743.56	7747
Increase over the period	125.71	151.99	120.91

Source : Directorate of Economics and Statistics, Ministry of Agriculture, Govt. of India, 2008.

Coconut Production in Tamil Nadu

Coconut cultivation in Tamil Nadu has history of over 150 years. One hundred year old trees can be seen even today in some of the villages. In the early periods, farmers did not consider coconut as a commercial crop. Coconut palms were grown along the periphery of the field crops and the products were utilized mainly for social and religious purposes. The reasons for taking up commercial coconut cultivation over the past 50 years are assured source of income, increased cost of cultivation of traditional field crops, dearth of labour force, changed work culture among the village youth. Many farmers in Tamil Nadu feel that they were forced to take up coconut farming as their standing crops were affected by the shade of coconut trees from neighbouring farms. Coconut cultivation was an alternative strategy to earn income from the farms when field crops can not be raised for want of adequate water for irrigation. Coconut was identified as only crop, which can survive all through the year with little quantity of water. Coconut plantations were generally free from major pests and disease attack till recently (Subburaj, 2000).

Coconut Situation in Tamil Nadu

In spite of the excellent investment opportunities offered by the crop, Tamil Nadu is lagging behind in the field of production of diversified coconut products and coconut based agro industries. Production of coconut products particularly non-conventional coconut products such as coconut milk and powder, coconut cream, coconut gel, vinegar, processed tender coconut water, coconut chips etc., are yet to be initiated. As of now, the coconut processing sector in Tamil Nadu is largely represented by coconut oil mills / industries and a few coconut desiccated units and the non-kernel based coconut industries such as coconut fibre and coir units. The other by-products such as timber, leaves, midribbes etc., are malused, although they can be converted into value added products having high export potential.

At present, there are about 500 coconut oil mills (small and big, either with or without expeller) in Tamil Nadu out of which, 210 are located in Kangayam region only. Nearly 60 per cent of the oil production in Tamil Nadu is contributed by the Kangayam region while the remaining per cent is by Thoothukudi, Peravoorani, Nagarcoil, and Thanjavur regions. It is estimated that the mills at Kangayam region have a crushing capacity of about 1.5 lakhs tones per day per shift per annum and nearly 75 tones copra are produced per day (Subburaj, 1997).

As for production of desiccated coconut powder, there are 17 units in Tamil Nadu with an annual production capacity of nearly 6000 metric tones and only one unit has taken-up production of coconut milk powder. Since quality nuts are needed for desiccation, coconuts produced in Pollachi, Peeravoorani, Thengonikotta, Batlagundu and Nagarcoil regions only are used. Although, desiccated powder and milk do not find better markets within Tamil Nadu, they are moved to upcountry markets and foreign markets through wholesalers.

As for tender coconuts, harvest of tender coconuts, as a strategy is not common among coconut growers; and it is estimated that less than 16 per cent of coconut growers harvest tender coconuts. Yet the demand driven marketing logistics for tender coconut is found particularly in urban centres and motels on national highways in Tamil Nadu. However, the market share of tender coconuts in soft drinks industry is very low.

As for institutions promoting marketing of coconuts, the Regulated markets play an important role. Out of 272 Regulated markets in Tamil Nadu, 27 are exclusively dealing with coconut and coconut products (copra) due to development of commercial markets (secondary markets) for coconut products in a few geographical clusters and among them, 15 are operating in Erode district. In fact, they are the only organised marketing institutions dealing with coconut and coconut products while, no cooperative marketing society is involved (Subburaj, 2004).

Trend in Area, Production and productivity of coconut in Tamil Nadu

The state recorded an increase in cultivated area under coconut from 0.66 million ha during 1950-51 to 383 million ha during 2007-2008, during the 50 year period. The share of the State in the area, under coconut has gone up from 10.54 per cent of the national area during 1950-51 to 20.14 per cent during 2007-2008

(CDB, 2001). This would mean that the increase in the area under cultivation has been going up at a more rapid rate than it was in many other states including Kerala and the National average.

Table 3.8
Area and Production of Coconut in Tamil Nadu since 1950-51

Year	*Area ('000 ha.)*	*Production (million nuts)*	*Productivity (Nuts/ha.)*
1950-51	66.00	462.60	7009
1955-56	51.50	417.30	8102
1960-61	55.00	445.30	8147
1965-66	90.20	843.60	9352
1970-71	95.10	941.50	9900
1975-76	109.90	1098.60	9996
1980-81	116.00	1354.40	11675
1985-86	149.60	1494.50	9989
1990-91	226.40	2358.30	10416
1995-96	298.60	4345.70	14553
2000-01	323.50	3158.40	9763
2005-06	370.60	4867.10	13133
2006-07	374.60	5429.90	14495
2007-08	383.37	4968.20	12959
% of increase over 1950	580.86	1073.97	184.89

Source: Directorate of Economics and Statistics, Ministry of Agriculture, Govt. of Tamil Nadu, 2008.

The production of coconut in Tamil Nadu has gone up from 462 million nuts during 1950-51 to 4968.20 million nuts during 2007-2008 registering a 1073 per cent of increase during the 57 year period. The share of Tamil Nadu's production to national coconut production has gone up from 14.08 per cent during 1950-51 to 33.70 per cent during 2007-2008. The increase in production in the State had been far above the national average increase (CDB, 2003). Tamil Nadu has also made impressive gains in productivity. The productivity of coconut in the state has gone up from 7,009 nuts per ha during 1950-51 to 12,959 nuts per ha during 2007-2008 which is far above the national average of 7,608 nuts per ha.

Use of Coconut Palms

The early adopters of coconut palms had used coconuts including tender nuts for religious and domestic purposes. Use of coconut oil was not uncommon. Byproducts such as stump, frond, etc., were used for fuel and other domestic purposes. They were given to neighbours free of cost. However, coconuts were sold to neighbours as also in the local markets. Since coconuts and by-products were abundantly available due to increase in coconut palms, farmers with coconut palms in groves and in large numbers brought their dehusked coconuts to local market places/urban centers for sale. Itinerant traders were also attracted and they purchased coconuts in villages. Cottage and village industrial fibre units were also developed and they purchased coconut husks available with farmers. The origin of the practice of sale of coconuts to channel intermediaries could be thus traced to 50 years from now which still continues as such.

Domestic Demand of Coconut

Even though the coconut cultivation is concentrated in the coastal tracts in the country, the demand for coconut and its products exists throughout the country. Coconut is widely used for culinary and religious uses in India besides processing it into edible and milling copra, desiccated coconut, etc., mainly in the coconut growing states. Besides, coconut in tender stage is widely used in the country as source of nutritious thirst quenching drink. While coconut kernel is the sine qua non of the dietary of the people of the West Coast of India, it is an essential ingredient in many sweet dishes and culinary preparations in other parts of the country. Milling copra, from which coconut oil is extracted is the major coconut product of commerce in the country. Even though coconut oil is used for edible, toiletry and industrial purposes, the toiletry segment absorbs roughly 50 per cent of the production in the country. Coconut oil as a hair oil is also used throughout the country like matured coconut and tender coconut. While the demand for coconut oil as an edible oil and industrial oil is influenced by its price, the demand for coconut oil as hair oil is not influenced by its price. The demand for coconut oil as a dietary fat is mainly confined to Kerala, the major coconut growing state in the country and border districts, Mangalore in Karnataka and Kanniyakumari in Tamil Nadu. The demand for coconut oil as an edible oil is also influenced by the availability of other substitute oils, especially low priced imported palm oil. Therefore coconut oil as cooking oil is vulnerable to substitution when it maintains premium price.

With the availability of low priced vegetable oils and petroleum based derivatives, the use of coconut oil and coconut oil derivatives for industrial uses has also decreased significantly during the last decade. For example, while the soap industry had consumed about 25 per cent of the total coconut production during the 1950's, the industry consumes only about 5 per cent of the total production of coconut oil at present. Due to the high price of raw materials, coco-chemicals (derivatives from coconut oil) are not manufactured from coconut oil to any significant level.

In the olden days, Indian coconut oil was popular in Europe and America for its quality and aroma. It was known as "Kochi oil" because coconut oil from South India was mainly exported from the Port of Cochin. Unfortunately, India became an importer of coconut oil later and the popularity of Kochi oil gradually faded. Still, Indian coconut oil is considered the best in the world since it is made from quality copra. The availability of quality copra has increased in the country due to modern processing technologies. Recent years have also witnessed the marketing of large number of branded coconut oil in small packs in and outside the country. Value addition to coconut oil can be done by making it medicated for using as hair oil and body oil, which has demand throughout the country and in Middle East. India is exporting coconut oil in small quantities to many countries in spite of its price disadvantage. Desiccated coconut (DC), coconut milk and spray dried coconut milk powder are the kernel based convenience products manufactured in the country. Except DC, market for other products is insignificant. However, spray dried coconut milk powder is exported in small quantities. Even though coconut milk is used in many recipes in the West Coast of India and to make Thai foods in hotels, majority of the households use coconut milk manufactured in the traditional way. DC is widely used to manufacture confectionery and bakery products. Its use in households is very limited. It is also exported on a limited scale from India.

Packed tender coconut water, concentrated coconut water and coconut water based vinegar are the products that have entered the market recently. There is an increasing demand for natural vinegar made from coconut water in the developed countries on health ground and the present production is mainly consumed by the domestic market.

Coconut shells charcoal, shell powder and coconut shell based activated carbon are the other commercial products of importance. Shell charcoal is the raw material for the manufacture of shell based activated carbon. It is exported to Bhutan, Oman, Singapore, Sri Lanka, Kenya, The Netherlands, Italy, Malaysia, Mauritius,

Japan, USA, UAE and UK. Shell powder has domestic demand for use in mosquito repellant coils / mats, agarbathis, plastic and bakelite industries. Coconut shell is used to make various utility articles and show pieces, which have domestic market. Shell ice cream cups have export market and India has already exported substantial quantity of shell ice-cream cups to Europe. Coconut oil cake has market throughout the country as cattle feed (Srivastava, 2002).

The fibrous husk covering the inner shell called coco fibre is the raw material for making coir. It is a versatile hard fibre used commercially for the manufacture of coir yarn, mat, matting, carpet, geo-textiles, rope etc., coir industry is a traditional cottage industry earning foreign exchange besides the means of livelihood for more than 10 million peoples (Markose, 1994).

Coir pith is the waste material when coir fibre is processed, which could retain moisture by eight times of its weight. Studies by Kadalli and Suseela Nair (2000) found that its moisture retention property is around 500-600 per cent higher than potassium and micronutrient contents. Coir pith is also converted into organic manure. It also contains around 77 per cent volatile solids and therefore, used as a cooking media by producing biogas (Mathew et. al., 2003). Being a bulk material for transport, technology has also been developed to compress it into coir briquettes of various shapes by reducing the volume to about five to eight times of the whole mass. Coir geo-textiles and coir pith are promising products of coconut industry in the coming years (Coir Board, 2000).

Consumption Pattern of Coconut in India

Coconut is widely used in India for culinary uses and religious purposes besides/processing into products like milling copra, edible copra, coconut oil, desiccated coconut, etc. While 50 per cent of the production is used for culinary and drinking purpose in raw form, only 35-40 per cent is used for -the manufacture of commercial products, 8-10 per cent is used as tender nut form and rest of the 2 per cent is used by the coconut farmers for their seed nut. Of the 35-40 per cent utilized for commercial purposes, roughly 30 per cent is used for manufacturing milling copra to extract oil. The remaining is used for making edible copra and convenience food products. Of the coconut oil produced in the country, 30-40 per cent is used for edible purposes, 40-50 per cent for toiletry purposes mainly as hair oil and the remaining for industrial purposes. Of the 50 per cent of raw nut 95 per cent of the raw nut is used for domestic purposes and rest of the 5 per cent is used for industrial purposes. Desiccated coconut, coconut cream and coconut milk powders etc. are manufactured out of the industrial purposes. About 8-10 per cent of the production is utilized as tender nut for drinking purpose in the country. Of the coconut husk available in the country, only 25 per cent is used for manufacturing coir and coir products and the remaining is mainly used as fuel. Coconut shell is mainly used as a domestic and industrial fuel and only a small quantity is used for making charcoal, shell powder and handicrafts. The water from matured nut is generally wasted and only insignificant quantity is used for making commercial products such as beverage, vinegar, etc. Coconut water is also used for the production of bio-gas in Thiruvambady coconut society. Coconut oil cake is traditionally used as cattle feed. Nowadays it is mainly used as raw material for extracting the remaining oil by solvent extraction process. The residue cake is used as cattle feed either as such or as fortified cattle feed incorporated with other ingredients (Kaul, 1994; Markose, 1994 and Aravindakshan, 1996).

Several studies have been conducted on the consumption pattern of coconut products in different states. For instance, Kerala consumes around 52 per cent of the production for edible and religious purposes, 30-31 per cent processed into milling copra, which is crushed into coconut oil, 10-11 per cent has been consumed as tender coconuts, 8 per cent as production of edible ball copra and the balance for the manufacture of non-traditional coconut products like desiccated coconut, coconut cream, milk etc. Tamil Nadu consumes around 47 per cent of total production of coconuts for milling copra as well as desiccated coconut and around 10 per cent as tender nuts, the remaining for raw consumption and traded outside the state, mostly in northern and western India (Venketachalam and Thomas, 1997, Subburaj B. 1996, 2000). It is estimated that around 25-40 per cent of the production in Karnataka and Andhra Pradesh is consumed in the form of tender nuts and the rest for production of ball copra, besides the use in raw form, about 60 per cent of the production in Orissa and 80 per cent in West Bengal are marketed as tender coconuts (CDB, 2001). The consumption pattern of the Andaman and Nicobar Islands, which is a traditional coconut belt, differs between the islands. It is estimated that about 42 per cent of the coconut production in the islands are consumed for edible purposes, 30 per cent for milling copra and 28 per cent as tender coconuts (Thomas, 2002).

Chart 3.1
Consumption Pattern of Coconut in India

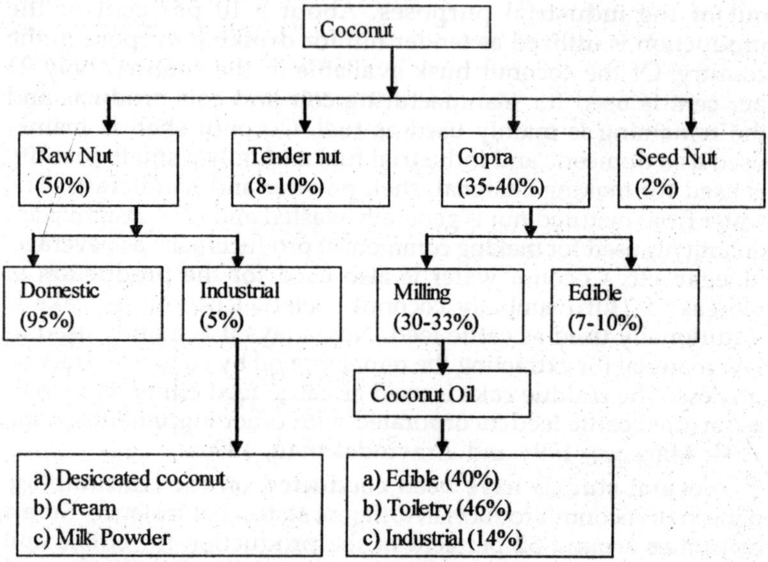

Source: Coconut Development Board, India.

Export of Coconut and its Products

India's export earnings from coconut kernel products are comparatively insignificant. The country gets major export earnings from coir and coir products. Export value of coir and coir products from India during the year 2000-2001 was Rs. 3,136.6 million. Foreign exchange earnings from other products were Rs.274.2 million only. Among the other products, coconut oil was the single largest commodity exported followed by desiccated coconut. The country is still pinning its hope to increase its export earnings mainly from coir, the demand of which is on an upward trend as an eco-friendly product. The coir products exported include coif fibre, coir yarn, coir rope, curled coir, handloom matting, power loom matting, tufted mat, power loom mat, handloom mat, geo textiles, rubberised coir, etc. Coir pith and coir geo- textiles are the two non-traditional products that have export potentials market. Coir pith is in demand for soil conditioning and mulching. In the recent past the coir pith is gaining popularity as a growing medium for horticultural crops.

Coir geo-textiles are gaining wider acceptability in soil erosion control.

A major constraint for export of Indian coconut products, mainly copra and coconut oil, is the wide gap existing between the domestic and international prices. On an average, the domestic price is higher by 40-50 per cent than the international price (Srivastava, 2002).

Import of Coconut Products

The main products imported by different countries are copra, coconut oil, desiccated coconut, oil cake, and coir products. The European and American countries are the major importers of coconut oil. During 1998, European countries imported 51 per cent of the total export volume of coconut oil followed by America and Asian countries with the share of 33 and 14 per cent respectively. Import of copra was quite insignificant by these countries. Recently copra meal also figures prominently among the imported goods. The import of desiccated coconut was almost uniform among all the major countries probably because of its extensive application in confectionery industry. Coir and coir products also enjoy a significant intake by the American and European countries. Under the WTO regime, more coconut products are likely to enter the foreign markets and improve the import scenario as well which would indirectly benefit the global coconut industry (*Rethinam, 2001*). Coconut oil refined and oil cake are the major items of import into India from Indonesia, Philippines, Malaysia, Singapore, UAE, Sri Lanka and Nepal.

Coconut Food Products

Coconut is food, beverage and oil seed. Coconut in tender form is consumed as a beverage and the kernel of matured coconut forms an ingredient in many food preparations. Copra, the dried kernel is the richest source of cooking oil and the edible copra is a popular dry fruit in India.

Coconut usage is important in the therapeutical sector especially in the Indian system of medicine. Coconut kernel, coconut milk, coconut oil, coconut inflorescence, cabbage (tender shoot at the growing point), tender coconut, coconut water, toddy (palm sap) and coconut jaggery possess immense medicinal properties and are used in various ayurvedic preparations.

Inflorescence, tender coconut and unfermented sweet toddy

are rich sources of sugars, minerals, essential amino acids, vitamins and digestive enzymes. Coconut milk is a good source of protein and fat. Coconut oil contains capric, myristic and stearing acids in addition to glycerides and is considered as the safest vegetable oil for human consumption.

Coconut Water

The water of tender coconut (TCW) is sterile, nutritious and thirst quenching health drink. It possesses therapeutic properties. The TCW has a calorific value of 17.4 per 100gm. Mineral Composition (mg per100 ml) is Potassium 312, Calcium 29, Magnesium 30, Sodium 105, Phosphorous 37, Sulphur 24, Chlorine 183, Iron 0.10 and Copper 0.04 (Coconut Development Board, 2006)

TCW is rich in potassium and other minerals. Sugars form an important constituent of the TCW. The concentration of sugars in the nut water steadily increases from about 1.5 percent to 5-5.5 percent in the early months of maturation and then slowly falls reaching about 2 per cent at the stage of the full maturity of the nut. In the early stages of maturity, sugars are in the forms of glucose and fructose (reducing sugars) and sucrose (non-reducing sugar) appears only in later stages which increases with the maturity while the reducing sugars fall.

TCW contains vitamin C and vitamins of B group. The concentration of Vitamin C ranges from 2.26 to 3.7 mg per 100ml. The drink is at its optimum level of acceptability when the nuts are of 6-7 months of maturity. TCW is a good rehydration medium for patients suffering from gastroenteritis and intestinal disturbances.

It is sweet, cooling, digestion inducing and an internal cleaning agent of human body system. 'Nilekerasavam' is a preparation from coconut water along with other herbal products. It is recommended for curing impotency. It is also rejuvenating, improves vitality and is gerentologically important as it regards the ageing process.

Snow Ball Tender Nut

Snow ball tender is the white round soft kernel pulled out from the shell with the water intact using a flexible knife. 8-9 months old nuts are used for the purpose. The CPCRI under a sponsored research project of the Coconut Development Board has recently developed the technology for preparing snow ball tender coconut. It is known as egg coconut in Malaysia.

Coconut Water/Beverages

Tender coconut water can be preserved and packed in aluminium cans and retortable pouches for marketing as a soft drink. Concentrated coconut water is also manufactured and marketed in the country. The product is served as a soft drink water diluting with water. Carbonated coconut water soda is also made from coconut water concentrate.

Tender coconut water is highly effective against dehydration due to cholera and dysentery. Tender mango leaves crushed in tender coconut water is administered to jaundice patients. 'Elaneer Kuzhambu' is a popular ayurvedic preparation, extensively used as eye lotion for general eye diseases, especially against cataract.

Preservation & Packing of Tender Coconut Water

Tender Coconut Water The water of Tender Coconut is a sterile, nutritious, thirst quenching health drink. The pure and sweet tender coconut possesses enormous therapeutic properties and is a thirst quenching refreshing health drink for all age groups. It is loaded with natural sugars, salts, and vitamins and is low in fat, calories and has no cholesterol. It's a highly rejuvenating tonic that can ward off fatigue in an instant. Tender coconut water has the same level of electrolytic balance as we see in our blood. The processing and packing of tender coconut water with a capacity of 10,000 tender nuts per day involves an investment ranging from Rs.35-40 lakhs covering plant and machinery and margin money for working capital with a direct employment potential of 30 personnel. The profitability after taking into account the prevailing prices of finished product works out to be around 20 per cent with a pay back period of 3 years. Tender coconut water enjoys a great potential as a health drink in India and international market. The Coconut Development Board in collaboration with Defence Food Research Laboratory, Mysore has developed the technology for packing tender coconut water in pouches/aluminium cans, which is available to entrepreneurs at a total lumpsum transfer fee of Rs.3 lakhs.

Process

Coconut Water of 6-7 month stage is first filtered through pressure filters and then mixed with the desired proportion of additives plus sugar and concentrated to the appropriate level. The water is then packed in pouches/cans and retorted in an autoclave, after which it is cooled in a stream of cold water.

Coconut Vinegar

Vinegar is manufactured from matured coconut water. Vinegar is used in pickle industry, salads, sauces and various condiment preparations as a preservative and flavouring agent. Vinegar aids digestion and improves the quality of cooked meat and fish.

Nata-De-Coco

Nata-de-coco a cellulosic white to creamy yellow substance formed by acetobacter aceti subspecies Xylinium, on the surface of sugar enriched coconut water / coconut milk / plant extract / fruit juices or other waste materials rich in sugar. It is another product manufactured from coconut water. It is popularly used as a dessert. It is also used as an ingredient in other food products, such as ice cream, fruit cocktails and frozen delights. It is also packed into a delicious cream pie and mixed with beverages, wine and liquor. It is a gelatinous product formed by the action of the microorganism. Acetobacter xylinium in a culture medium of sugar enriched coconut water. Technologies for the preservation and packing of TCW and manufacture of vinegar and nata-de-coco from coconut water are available with the Board.

Process (from matured coconut water)

Initially coconut water is strained and mixed with sugar and glacial acetic acid in stipulated proportions. Boil for ten minutes and cool. Add the culture solution and distribute the mixture in wide mouthed glass or plastic jars, cover the jar with a paper or a thin cloth to protect from dust. It is then kept aside undisturbed for two to three weeks. After this period, the white jelly like thick surface growth is harvested, washed thoroughly to remove all the acids and sliced into cubes. It is then immersed in flavoured sugar solution, again boiled and packed in glass jars or retortable pouches, sterilized and sealed.

Coconut Kernel

Kernel of coconut is nutritious and used as an ingredient in many food preparations. Composition of coconut kernel is Moisture 45%, Fat 37%, Protein 4%, Sugars7%, Fibre 3% and minerals 0.8%.

The major protein in the kernel is globulin which contains high concentration of glutamic acid (19.07%) and arginine (15.9%). Other amino acids present in appreciable amounts are lysine (5.8%), leucine (5.96%), proline (5.54%) and aspartic acid (5.12%).

It has a much higher content of arginine in relation of lysine; the lysine / arginine ratio being 0.36. The fresh kernel contains 15 IU of thiamine, 1 mg of ascorbic acid, traces of Vitamin A and 0.2 mg of tocopherol per 100 mg. It contains potassium 436 mg, calcium 13 mg, magnesium 52 mg, iron 2.08 mg, copper 0.32 mg, phosphorous 94 mg, sulphur 33 mg and chlorine 144 mg per 100 gm (Coconut Development Board, 2006). Studies conducted by the Department of Biochemistry, University of Kerala have shown that the dietary fibre and protein of coconut kernel have significant cholesterol lowering action.

Desiccated coconut, coconut milk, coconut cream, spray dried coconut milk powder and coconut chips are the convenience coconut products manufactured from the raw coconut kernel.

It helps to increase the body weight and is also an internal body cleaner. Daily use of coconut kernel is advised for overcoming ill health and impotency.

Virgin Coconut Oil

Coconut oil obtained from coconut milk is called virgin coconut oil. Traditional and modern methods are available for the manufacture of virgin coconut oil. In the traditional method the milk extracted from grated coconut kernel is boiled to get oil. Of late, the traditional method has been partially mechanized using a bridge press and mechanical grater. The modern method of extracting oil from fresh coconut kernel is known as wet processing.

The virgin coconut oil is considered superior for use as edible oil, hair oil and baby oil because of its pleasing aroma and purity. It is applied on the body of babies to protect from skin troubles. Because of its low FFA content, this oil has a long shelf life.

Desiccated Coconut

Coconut has been the staple food of people in many parts of Asia and Pacific Islands. The dehydrated shredded flesh of coconut known as desiccated coconut. A large number of units in India are manufacturing DC which is mainly absorbed by the confectionary and other food industries. DC is also used as substitute to grated coconut in various household preparations. DC is available in different grades based on the fineness of the material. It is often used as a substitute to grated coconut in food preparations such as curries, cakes, sweets and chutneys. Confectionery and bakery units are the main consumers of desiccated coconut. Desiccated Coconut Powder is obtained by drying ground or shredded coconut kernel after the removal of brown testa. It finds extensive use in confectioneries, puddings and many other food preparations as a substitute to raw grated coconut. In India the product is manufactured by small-scale units scattered in Karnataka, Tamil Nadu, Kerala and Andhra Pradesh. A study conducted by the Coconut Development Board has revealed that a growing consumer demand for desiccated coconut powder could be developed in the country by resorting to organised market promotion activities for the popularisation of the product in consumer packs for household uses. The survey has also shown that desiccated coconut powder in consumer packs is acceptable not only in non-coconut producing states but also in Kerala. From the survey it was revealed that a sizeable section of the middle class and upper class families residing in cities and towns in Kerala would prefer desiccated coconut powder, if readily available, to raw nuts.

Process

The fresh matured coconuts are dehusked and deshelled manually using hand tools. The deshelled coconut kernels in the form of round balls are pared using scrapers to remove the testa. The pared kernel balls are then cut open to drain off the water and then washed thoroughly in fresh water to remove the invert sugars from the inner surface of the kernel. The kernel is then ground into a fine mass using hammer or pin mills. The ground mass is blanched with live steam for about 20 minutes to bring down the microbiological counts. The blanched mass is then dried in hot air drier at a temperature of 80-90°C for about 10 hours so as to bring down the moisture content to below 3 per cent. The hot air drier is provided with a drying chamber consisting of a series of trays, which hold the feed. Hot air is blown into this chamber from an external source through blowers. The dried mass is tested for moisture, free fatty acid and microbiological counts. The product is packed in polythene pouches.

Coconut Cream

Coconut cream is the processed milk extracted from fresh matured coconuts. It is a ready-to-use product which can either be used directly or diluted with water in various edible preparations such as curries, sweets, desserts, puddings, etc.. Coconut cream when partially defatted is called coconut milk.

Coconut cream/milk is used as an ingredient in household recipes and as a component of processed foods. It can also be used in the manufacture of bakery products and for flavouring food stuffs. Processed and packed coconut cream has a shelf life of six months and once opened it should be stored in refrigerator for subsequent use. The technology has been developed by the RRL, Thiruvananthapuram under a sponsored project of the Board.

Coconut milk is also used as a mixer in alcoholic drinks. Coconut milk / cream is available in pouches, bottles and tetra packs. Technology for the manufacture of coconut cream is available with the Board.

Process

The first step is breaking the dehusked nuts into halves. The split nuts are deshelled to separate the kernel. These two operations usually are done manually. Kernel is washed and then blanched by immersing in hot water at 80oC for 10 minutes. The next step is comminution of kernel into small gratings using a hammer mill. The gratings are subjected to pressing using, continuous screw press to extract the milk. The coconut milk thus obtained is filtered by passing through a vibratory screen. Food additives such as emulsifiers and stabilizers, are to be added to the milk to obtain a stable consistency and texture. For this purpose, permitted emulsifiers and stabilizers are mixed with hot water separately and mixed thoroughly. This is added to the coconut milk and then subjected to emulsification using a mechanical impeller emulsifier. The emulsified milk assumes a creamy consistency. The coconut cream is then pasteurized at 95oC for 10 minutes in a plate heat exchanger. The pasteurized coconut cream is hot filled in cans using a mechanical volumetric filling machine followed by steam exhausting. The cans are seamed using an automatic can seamer. The seamed cans are sterilized in a rotary retort at 15 psi for 20 minutes. The cans are then cooled in running water. The residue obtained after the extraction of milk is dried in a hot air dryer to 3 per cent moisture level. The residue forms a base for making coconut burfi, coconut cookies, curry and chutney powders. Other by-products like coconut water and coconut shell could be processed into value added products. Vinegar and Nata-de-coco can be manufactured from coconut water. Shell charcoal and shell powder can be manufactured from coconut shells. The utilization of byproducts would improve the economic feasibility of the process.

Spray Dried Coconut Milk Powder

Spray drying is the best method for the preservation of coconut milk. Spray dried coconut milk powder is reconstituted into coconut milk by the addition of water which can be used to make various food preparations. The product offers additional advantages such as less storage space, bulk packaging at reduced cost and longer shelf life. Technology for the manufacture of spray dried coconut milk powder is available with the Board. Using coconut milk is not a cumbersome or time consuming process anymore for those who want to include this wholesome nut in their regular diet. The spray dried coconut milk powder is now available in convenient and ready to use packs with same freshness of a fresh coconut milk. It has a longer shelf life and is convenient to use. This can be used in place of fresh coconut milk for food preparations / beverages in households and food industries by dissolving it in water. Central Food Technological Research Institute, Mysore with the financial assistance of the Board has developed the technology for spray dried coconut milk powder which is available to entrepreneurs at a total know-how fee of Rs.5 lakhs. The total capital investment for plant and machinery along with working capital margin comes to around Rs.2.36 crores. The internal rate of return works out to be about 16 per cent with a pay back period of about 3 years. On an average one thousand coconuts would yield about 99.6 kg of spray dried coconut milk powder. Spray dried coconut milk powder has a tremendous market potential in India and in international market. Direct employment potential is around 40 personnel.

Process

The first step is breaking the dehusked nuts into halves. The split nuts are deshelled to separate the kernel. These two operations usually are done manually. Kernel is washed and then blanched by immersing in hot water at 80oC for 10 minutes. The next step is comminution of kernel into small gratings using a hammer mill. The gratings are subjected to pressing using, continuous screw press to extract the milk. The coconut milk thus obtained is filtered by passing through a vibrating screen. The coconut milk is then homogenized and mixed with maltodextrins and other emulsifiers in an additive mixing tank. The milk mixed with additives is then spray dried into a fine milk powder using a spray drier. The product is then packed in Aluminum foil packets in various sizes as per customer requirement.

Coconut Jam

Coconut jam is prepared from coconut skin milk by adding sugar and glucose. It is an ideal spread for bread.

Coconut Chips

Coconut chips' is ready-to-eat snack food. It is prepared in salted and sweetened forms. The Central Plantation Crops Research Institute, Kasargod has standardized the process for preparation of chips. Coconuts of 9-10 months old are used for the preparation of chips.

Copra

Two types of copra, namely the milling and edible, are made in India. The milling copra is used to extract oil while the edible copra is consumed as a dry fruit. Edible copra is made in the forms of balls and cups. Copra contains the highest percentage of oil compared to other oil seeds. It contains 15-20 percent carbohydrates, 9 percent protein and 4.10 percent crude fibre besides 65-68 percent fat. The carbohydrate fraction consists of a large percentage of cellulose and sucrose together with other sugars. Besides glucose and fructose the presence of galactose, raffinose and pentoses has been reported.

Coconut Oil

Coconut oil is rather unique cooking oil as it contains the short and the medium chain saturated fatty acids as the major component fatty acids. It finds extensive use in the food industry due to its characteristics such as easy melting behaviour, resistance to oxidative rancidity, pleasing flavour and good digestibility. Coconut oil has gained importance as a dietary fat because of its lauric acid, the source of monoalurin in the body and low content of Omega 6. It can be used for manufacturing margarine and shoternings. Coconut oil is preferred as a fat in the preparation of filled milk, infant milk powder, ice-cream and confectionary and bakery products. Because of its stable character, coconut oil is the preferred fat for deep frying.

Studies undertaken by the Biochemistry Department, University of Kerala showed that coconut oil: does not elevate blood total cholesterol, Increases blood HDL cholesterol, if consumed along with coconut kernel, lowers blood cholesterol, does not elevate LDL cholesterol or LDL cholesterol/HDL cholesterol ratio, decreases serium triglycerides.

A large number of branded coconut oil in consumer packs are marked in the country. Coconut oil is refined for industrial use. The refined coconut oil is water white in colour with no aroma. Since coconut oil has the lowest level of unsaturated fatty acids among all vegetable oils, the cost of hydrogenation is much less than all other oils.

A variety of ayurvedic medicines are prepared using coconut oil. 'Chandrakalalepam' (head ache), 'Dhavalalepam' (head ache), 'Sinduradi Lepam' (leprosy and veneral diseases), 'Karpooradi Thailam' (asthma and rheumatic arthritis), 'Gandaka Thailam' (leprosy) and "Murivenna' (bone fracture) are only few of them. The extract of 'Dhendappala Leaves' in coconut oil is recommended for soriasis. 'Karkokil' seeds fried in coconut oil is effective against leucoderma. Pineapple juice distilled with equal quantity of coconut oil is used against whooping cough.

Toddy

Fresh toddy obtained from tapping the inflorescence of coconut palm is an excellent beverage and a rich source of sugars (15-16%). It contains fair amounts of vitamins B1, B2, A and C. 17 amino acids are reported to be present in fresh coconut toddy, glutamic acid being the major one followed by thioneine and aspartic acid (Coconut Development Board, 2006).

It is a highly nutritious and refreshing drink, considered as the next best alternative for infants in the absence of breast milk. Sweet toddy helps to prevent internal blood coating and is also diuretic. Along with Glycyrrhiza glabra, it is effective against typhoid. Cyprus rhizomes ground in sweet toddy is administered to babies in case of stomach trouble. A multi-purpose high valued ayurvedic medicine – *'Vrana Bhairava Thailam'* used against typhoid, epilepsy, neurosis, skin diseases, arthritis etc. is prepared using sweet toddy as one of the components.

Coconut Jaggery

Fresh toddy when boiled to 118-120°C and allowed to cool solidifies. The solid mass is known as coconut jaggery or 'gur'. Coconut jaggery is made in traditional coconut growing tracts in the country on a cottage scale. Calcium and phosphorous are the important minerals contained in coconut jaggery.

Treacle is another product manufactured from sweet toddy. It is obtained by boiling down the toddy. Fresh toddy is also a good source of baker's yeast. The fresh toddy rapidly ferments and the sugar is replaced by about 5-8 per cent alcohol, which on distillation yields arrack. Fermented toddy on acetic fermentation yields vinegar containing 4-7% acetic acid.

"Coconut is the best and the first rate brain forming food; curd and milk belonging to the second rate; vegetables, pulses and fruits to the third rate brain forming food respectively".

Jaggery is extensively used in the preparation of health improving tonics for women after child birth. Coconut cabbage in combination with jaggery is effective against gynaecological diseases. Regular use of black coffee prepared with jaggery helps to overcome intestinal complaints. Daily use of jaggery with onion promotes the growth of children.

Coconut Cabbage

The young meristematic shoot in the heart of the crown is known as the cabbage. It can counteract complaints due to 'vatha' and 'pitha' and is also an aphrodisiac. Coconut cabbage, crused and mixed with jaggery is effective against gynaecological problems such as excessive bleeding and leucorreha.

Coconut Inflorescence

Coconut inflorescence is used for curing urinary complaints, back pain and headache and is also very effective against gynaecological diseases. It is an ingredient of *'Thengin Pookula Oil"* which is recommended against head ache, burning sensation of eyes and skull. *'Thengin Pookula Rasayanam'* is yet another preparation from coconut inflorescence. It is virtually a health tonic for lactating mothers and also relieves back pain. Regular use of coconut inflorescence along with milk in the form of a past guarantees is permanent relief from urinary complaints. The juice extracted from coconut inflorescence mixed with honey is used in the treatment of gynaecological diseases. It is also an ingredient of *'Ellam Nishadi Lepa'* used for *'Vatha-raktha-dosha'*.

Coconut Husk

The ash of coconut husk mixed with sugar and tender coconut water cures gynaecological diseases. The juice extracted from the husk of tender coconut stops dysentery.

Coconut Milk

It is an ingredient of several ayurvedic preparations for skin and head diseases. Coconut milk along with powdered mahua seed kernel can be applied as a cream against skin diseases. It is also used for ayurvedic preparations like 'Thilamazhadi lepam', 'Nalikeramoolam' (skin diseases) and 'Neelibringadi oil' (for hair growth).

Coconut Cake

Coconut cake is the residue left after the extraction of oil from copra which is mainly used as a cattle feed. Coconut cake contains 4-5 per cent oil which is extracted by solvent extraction process. This oil is generally used for industrial purpose and de-oiled cake is used to make mixed cattle feed. There are a few such units in the country especially in Kerala.

Production Process of Coconut Shell based Products

Activated carbon

Coconut shell is used for manufacturing a variety of products of commercial importance including activated carbon. Shell is carbonized by using methods like pit method, drum method, destructive distillation etc. Shell based activated carbon is extensively used in the process of refining and bleaching of vegetable oils and chemical solutions, water purification, recovery of solvents, recovery of gold etc. It is used in gas masks and a wide range of filters for war gases and nuclear fall outs. Coconut shell based activated carbon units adopt steam activation process to produce good quality activated carbon Activated carbon is a non-graphite form of carbon which could be produced from any carbonaceous material such as coal, lignite, wood, paddy husk, coir pith, coconut shell, etc. Activated carbon manufactured from coconut shell is considered superior to those obtained from other sources mainly because of small macropores structure which renders it more effective for the adsorption of gas/vapour and for the removal of colour and odour of compounds. The activated carbon is extensively used in the refining and bleaching of vegetable oils and chemical solutions, water purification, recovery of solvents and other vapours, recovery of gold, in gas masks for protection against toxic gases, in filters for providing adequate protection against war gases/nuclear fall outs, etc. Steam activation and chemical activation are the two commonly used processes for the manufacture of activated carbon. However coconut shell based activated carbon units are adopting the steam activation process to produce good quality activated carbon.

Process (Steam Activation)

The process of activation is carried out in two stages. Firstly the coconut shell is converted into shell charcoal by carbonization process which is usually carried out in mud-pits, brick kilns and metallic portable kilns. The coconut shell charcoal is activated by reaction with steam at a temperature of 900°C -1100°C under controlled atmosphere in a rotary kiln. The reaction between steam and charcoal takes place at the internal surface area, creating more sites for adsorption. The temperature factor, in the process of activation is very important. Below 900°C the reaction becomes too slow and is very uneconomical. Above 1100°C the reaction becomes diffusion controlled and therefore takes place on the outer surface of the charcoal resulting in loss of charcoal.

Coconut Shell Charcoal

Shell Charcoal is an important product obtained from coconut shell. Shell charcoal is used widely as domestic and industrial fuel. It is also used by blacksmiths and goldsmiths and in laundries. Shell Charcoal is also used to produce activated carbon. Activated Carbon produced from coconut shell has certain specific advantages as the raw material can adsorb certain molecular species. Shell is carbonized by using methods like pit method, drum method, destructive distillation etc. The shell charcoal is the raw material required for the manufacture of activated carbon. The shell charcoal is manufactured by burning shells of fully matured nuts in limited supply of air sufficient only for carbonisation, but not for complete destruction. The output of charcoal in the traditional pith method is just below 30 per cent of the weight of the original shells. In India the average output in the traditional method has been found to be 35kg of charcoal from 1000 whole shells or about 30,000 whole shells yield 1 tonne of charcoal. Sometimes, especially when the processing is defective, the output is still lower and nearly 50,000 shells are required to produce one tonne of charcoal. To obtain good quality charcoal, fully dried, clean, mature shells should be used. Now several modern methods are in vogue for the production of charcoal. In the modern waste heat recovery unit the heat generated by the burning of coconut shells is used for drying copra and shell charcoal is obtained as by-product. A simple and efficient method adopted for the production of charcoal on cottage scale is given below.

Process (Drum Method)

A M.S. Drum kiln is used for carbonization of shells. The drum consists of three sets of six 1" dia holes provided at its bottom, middle and upper layers and a lid. A detachable chimney is provided which is installed on the lid after closing the drum. The manufacture of charcoal requires optimum carbonization of raw shells in a limited supply of air so that there is neither unburnt shell nor ash due to complete combustion. The steel drum is filled with raw shells after placing temporarily a four inch diameter wooden pole in the centre of the drum. The wooden pole is then removed, leaving a hollow space in the centre which allows the flow of smoke during carbonization. To start carbonization, a piece of burning rag is dropped to the bottom of the drum through the hollow space in the centre. When the fire is well underway, the lid with the chimney is placed into position and the upper and the middle sets of holes are closed. Carbonization which starts at the bottom progresses as it goes up as well as radially from the hollow space. When carbonization is complete in a particular zone, a persistent glow can be seen in all the six holes of a set. When the bottom most set of holes indicate this situation, the middle set of holes is opened and the bottom set closed. The stoppage of air flowing into the bottom region avoids over-carbonization in that region. The progressive carbonization results in reduction in volume of contents and therefore more shells are added from the top. When the middle region is carbonized well, the top set of holes is opened while the middle set is now closed. A further addition of raw shells is done to fill the volume reduction to maximize capacity for burning. When the top region is well carbonized, the top set of holes is also closed, resulting in complete stoppage of air inflow to the drum. The drum is then cooled for about eight hours after which the product is ready for discharge.

Coconut Shell Powder

Coconut shell powder is manufactured from matured coconut shells. The manufacture of coconut shell powder is not an organized industry in India. The product finds extensive use in plywood and laminated board industry as a phenolic extruder and as a filler in synthetic resin glues, mosquito coils and agarbathis. Coconut shell powder is preferred to other alternate materials available in the market such as bark powder, furfurol and peanut shell powder because of its uniformity in quality and chemical composition, better properties in respect of water absorption and resistance to fungal attack. The product is manufactured in sizes ranging from 80-200 mesh. Keeping in view of the vast industrial uses, the demand for coconut shell powder appears to be promising.

Process

Coconut shells free from contamination of coir pith, etc., are broken into small pieces and fed into a pulverizer. The powder from the pulverizer is fed into a cyclone and the parallel product is collected in bag filters. The shell powder is then fed into a vibrating sieving machine and packed according to mesh size requirements for various end uses. The rejects from the sieving machine can be recycled in the pulverizer for size reduction. The main requirements for consistent good quality of coconut shell powder, are proper selection of shell of proper stage of maturity and efficient machinery.

Coconut Wood based Products

The coconut wood because of its distinct grain characteristics is ideal for making wall panels, furniture, doors and windows, show pieces, etc. There are several small scale units manufacturing a variety of articles from coconut wood.

Coconut Leaves

Coconut leaves are plaited and used for thatching houses and sheds in rural areas. It is also used for thatching 'honeymoon huts' and such huts in town and cities. Technology is available with Regional Research Laboratory, Thiruvananthapuram, Kerala for extending the life of leaf thatch from one year to four years. Plaited coconut leaves are also used for making baskets, headgears and for erection of temporary fences. Plaiting of coconut leaves is a cottage industry in traditional coconut growing states. Midribs of leaves are used to make brooms of different types which are used for cleaning rough grounds and floors. Brooms of midribs of coconut leaves are manufactured on a commercial scale in Tamil Nadu and Karnataka.

Coir Pith

Coir pith a waste product obtained during the extraction of coir fibre from husk is very light, highly compressible and highly hygroscopic. It is used as a soil conditioner, surface mulch/ rooting medium and desiccant. Composted coir pith is an excellent organic manure for indoor plants as well as for horticulture crops. Several firms are manufacturing composted coir pith in the country. Compressed coir pith in the form of briquettes for easy transportation is also manufactured in the country.

Industrial Uses of Coconut Products

Shell, kernel, copra and oil derived from coconut are the important products that have to be used efficiently. The fresh kernels are nutritive for health due to proteins with essential amino acids and the oil. When sometime ago the coconut oil was considered to aggravate the cardio vascular disease due to its high content of saturated fats, the findings differ now. There is no scientific evidence to prove its harmful effects. In fact, the saturated fatty acids consist of medium chain fatty acids, which form a part of baby foods, weaning foods and other special preparations essential for good health.

Shell Products

Coconut shell, the hard woody layer directly covering the coconut meat or kernel, is multifariously used to manufacture products like charcoal, activated carbon, shell flour etc. The burning of coconut in a limited supply of air sufficient only for carbonisation, produces charcoal. Coconut shell charcoal, produced by the pit method, drum method of kiln method, reportedly contains the highest percentage of fixed carbon of all the ligneous charcoals.

The major use of this charcoal is as base material for activated carbon. This product has certain natural outstanding properties and is excellent for specific purposes. It is superior to other carbons, has more resistance to abrasion, higher capacity of absorption and higher purity in ash. This type of activated carbon has a small macropore structure, which makes it ideal for gas / vapour phase applications and for the removal of taste and other compounds. The activated carbon can be commercially produced by the steam activation or chemical activation process. Each method of production has its own advantages and disadvantages. Activated coconut shell carbon has good international market.

Being environment-friendly, attempts should be made to give an environmental designation to coconut products. Coconut oil, which is more popularly used in cosmetics, is a natural product of renewable resource. It can be efficiently promoted for the niche markets.

Coconut shell flour is clear, light brown, free-flowing powder, which results from pluverising of the shells of matured coconuts. It is produced in different mesh sizes, the most common of which is 100 mesh. This powder can be used as biodegradable filler in plastics as a substitute for sawdust and other fillers.

Copra and Desiccated Coconut

Copra is the dried coconut meat or kernel from which coconut oil is extracted with meal as its by-product. Depending on the process of extraction, copra is reported to produce from 57 to 64 per cent coconut oil. High-grade copra is obtained from matured, ungerminated nuts. Nuts intended for the production of copra are stored for 3-4 weeks after harvest before splitting. This practice enhances the quality of copra, as the moisture content of the fresh meat is reduced, thus making the meat thicker for oil, resistant to micro organisms and easier to detach from the shell during the subsequent drying process.

Desiccated coconut is the pure white, dehydrated and shredded meat of the coconut, which is prepared from the kernel. It is an excellent food product, popular in USA. Being one of the high value-added items, the industry has tremendous scope in India and abroad. The fat present in desiccated coconut is edible and highly digestible. The proteins have all the amino acids essential for human growth. Sugars are present at sucrose, glucose and galactose whereas carbohydrates are in the form of cellulose and pentose.

Cocochemicals

Cocochemicals are derivatives obtained from coconut oil. These chemicals can be divided into basic or primary groups, namely glycerol, fatty alcohols, methyl esters and soaps. The simple chemical process to produce these primary chemical derivatives is by hydrolysis (fat splitting). It is the treatment of coconut oil with high-pressure steam resulting in 3 mols of fatty acids and 1 mol of glycerol. Direct esterification of coconut oil with methanol gives methyl esters of coconut oil and glycerine. Hydrogenation of oil at high pressure and temperature in the presence of a catalyst gives a product suitable for bakery products.

Coconut Oil and its Derivatives

Some consumers prefer fresh coconut oil extracted from good quality copra, for its characteristic flavour and colour, while others prefer the bland oil obtained after refining and deodourising the raw oil. The main use of coconut oil, raw or refined, is for edible purpose particularly in South India. The bland oil remaining stable for long period of storage is preferred for use in food processing. The hardened coconut oil is preferred as a coating for cookies and candies.

The next important use of coconut oil is in the manufacture of soaps, since the lower fatty acids present in it give profuse lather. One can incorporate this oil up to 25 per cent along with other oils to obtain a good toilet soap. But with the high cost of coconut oil, one has to resort to the use of merely 5-10 per cent. Among the vegetable oils, coconut oil has maximum content of glycerine (13.5 per cent) which is premium price and finds good demand as natural glycerine for use in cosmetics. Hence, it is worthwhile to hydrolyse non-edible grade of coconut oil by splitting under high-pressure process, and fractionate the mixed fatty acids to recover the glycerine. In fact, the efficiency of recovery of glycerine by this process is more compared to the conventional method. In order to conserve coconut oil for edible purpose, one can substitute coconut oil in soap manufacture by Alpha Olefin Sulphonate (AOS), which is currently available in India. This substitute has additional advantages of improving the lime dispersing properties.

Major component fatty acid of coconut oil is lauric acid (50 per cent approx.) followed by myristic acid (20 percent approx.) both of which have wide industrial applications. On the other hand, the undesirable short chain voltalie acids of caproic / caprylic and capric (15 per cent approx.) which are skin irritants in soaps, can be fractionated and converted into value-added products like artificial fruit flavours, perfumes, chemical intermediates and plasticisers.

Coconut Oil in Cosmetics

Among the vegetable oils of Indian origin, coconut oil has the largest application in cosmetics. It is the oldest and most widely used because of its low incidence of irritation, sensitisation and for the simple good feel it gives on the human skin. The following information highlights the applications of coconut oil in cosmetics.

As Body Oil

Coconut oil, mostly preferred for massaging before bathing – especially in the case of newborn babies, is due to its low viscosity, pleasant aroma and easy washability. Moreover, the trend now is to substitute synthetic liquid paraffin oil with stabilised vegetable oil like coconut oil, which is more natural in relation to composition of the skin fats. Liquid paraffins are chemically unrelated to the skin fats. The age-old use of coconut oil as a body and hair oil is due to the unique unsaponifiable matter (0.1-0.4 percent) consisting of sterols, tocopherol, squalene and tritterpene alcohol. The thin film of coconut oil left on the body acts as a protector of the skin; thus, it is a unique natural vegetable oil suitable for cosmetic use.

As Hair Oil

The oil is also commonly used as hair oil. However, it has a tendency to become rancid on storage. This can be prevented by incorporation of anti-oxidants in small quantities. In order to attract consumers, the incorporation of fragrances came into vogue, which adds to the cost of the finished product. In order to prevent the oil from becoming rancid and also on account of its high price, mineral oils are admixtured with the coconut oil. These mineral oils are petroleum products with toxic aromatics that are eliminated by chemical treatment. These are then termed as 'liquid paraffins;' and permitted for use in cosmetics. They may be safe if properly purified and tested by sophisticated instruments, as it is found that on application these liquid paraffins get absorbed into the skin and may be harmful.

In campaigns, these are called as non-sticking hair oils by TV and other media. What good or harm it does to human body is to be judged by the user. Industry likes to use liquid paraffins due to its stable price, non-sticky nature and resistance to rancidity. Bureau of Indian Standards (BIS) has three types of hair oils: The first containing single or mixture of vegetable oil, the second containing only liquid paraffins and the third type allows mixture of vegetables oils and liquid paraffins

Hair oil manufacturers are taking undue advantage of the lacuna in standards by labelling as 'Coconut Hair Oil', which is actually a mixture of coconut oil and liquid paraffin. It is high time that BIS rectifies such anomaly.

Oil Based Surfactants

Copra has the highest oil content (around 65 per cent) as compared to other oil-bearing materials. The production of coconut oil in the last four years has increased. Consumers generally prefer expelled coconut oil for its natural characteristic flavour for edible purpose, while the solvent extracted oil is used for industrial purposes. But the high price of coconut oil compared to other oils, restricts its consumption for both edible as well as for industrial purposes.

Among the vegetable oils, coconut oil is unique as it contains maximum amount of saturated fatty acids, the predominant one being lauric acid. This when neutralised by sodium or potassium alkalies gives a soap that has excellent lathering, detergency and hardness properties. Thus, coconut oil forms an important raw material in the production of toilet soaps, liquid soaps, shaving cream and natural shampoo. All over the world, good quality toilet soap contains 20-25 percent of coconut oil and the rest being other hard fat. In India, mutton tallow imports are banned and substitute is hardened vegetable oil such as rice bran, castor, linseed or imported palm oil. It is interesting to note that due to high price of coconut oil, its use in toilet soaps is kept at minimum level resulting in soaps with poor lathering properties.

Many of the toilet soaps in India are currently made with only 50-60 percent of fatty matter against the minimum requirement of 78-79 percent of total fatty matter for export, on the pretext that India is trying to economize on the fatty raw materials, which have to be imported. Thus, one will find that the annual consumption of coconut oil in soaps in India has comparatively decreased due to reduced quantity, while the production of soap has increased mainly due to increase in population. This would mean less market demand for coconut oil, the main reason for being high cost of coconut oil. Even for edible purpose, consumers have switched over to other cheaper oils like palmolein.

The word 'surfactant' means surface active agent and soap falls under this classification as a natural surfactant, whereas the term 'surfactant' is generally used for synthetic surfactants that are derivatives of fatty acids, namely, sodium laurel sulphate and sodium lauryl ether sulphate. These constitute the active ingredients in a synthetic shampoo, which is fully biodegradable. Soap from coconut oil can be made by the cold process, the semi-boiled process and the full boiled process. The glycerin liberated during saponification remains in the soap and is wasted. While in the full-boiled process, the by-product glycerine is recovered by treatment with chemicals and the yield is poor. The modern trend is to split the coconut oil under high pressure by hydrolysis with water in an autoclave without any catalyst and recover the valuable glycerine, which, in turn has many applications. It is an important ingredient in both, powders and toothpaste.

Sodium lauryl sulphate has detergent powers far exceeding those of natural soap. Soaps are ineffective in hard water, sensitive to the presence of other ions. Hence synthetic detergents are superior and are effective in hard water and seawater. The other derivatives of lauric acid namely, sodium lauryl ether sulphate and sodium lauryl sarcosinate are all useful surfactants in industry, principally as emulsifiers, solubilisers, dispersants or detergents. They enable the formulator to make homogenous mixture, dispersion or emulsion of oily or waxy substance with water to disperse solids in liquids or disperse mutually insoluble liquids with other liquids. Creams and lotions are made more attractive in appearance, convenient to use and stable in storage, through the appropriate use of surfactants. Thus, many surfactants are obtained from coconut oil.

The higher fatty acid fraction (nearly 20 per cent) of coconut oil fatty acids, consisting of C16-C18 fatty acids, can be converted to alpha olefins by dehydration of the fatty alcohol fraction at elevated temperature with catalyst. The sulphonated alpha-olefin (AOS) is a good raw material for toilet soap, detergent and shampoo. Toilet soap with 10 per cent AOS is claimed to be better than that with 15 per cent coconut oil. Also AOS costs less than coconut oil. It has even lime soap has appeared in the market based on a combination of regular soap and AOS. The expensive coconut oil in the toilet soap base can be reduced or eliminated by the addition of AOS, which gives excellent flash foam. This is another reason for decrease in demand of coconut oil in toilet soap.

A recent development, which is not yet commercialised in India, is the use of methyl ester sulphonates from hydrogenated coconut oil by reaction with methyl alcohol. The glycerol can be recovered while the coconut methyl esters are sulphonated under optimum conditions and neutralised as sodium salt. This sodium methyl ester cocoate is a good surfactant, 100% biodegradable and relevant as scum dispersants. The high price of coconut oil and other techno-economic viability has prevented its implementation. It is a valuable substitute for linear alkyl benzene sulphonate (LABS) developed from renewable natural fats and oils. Another surfactant that has been developed from coconut oil is mono-ester formed by the reaction of lauric acid with sucrose. This mon-ester is soluble in many organic solvents, including ketones, alcohols and chlorinated hydrocarbons.

These sucrose mono-esters are useful as detergents, emulsifiers and wetting agents. They are natural. Being non-toxic, they can be used in cleaning food processing equipment, kitchen utensils, washing pesticides from fruits and vegetables as well as food emulsifiers and components of drugs and cosmetics. The diesters of sucrose are even more soluble in solvents and oils than monoesters. They can be used as emulsifiers for water-in-oil emulsion. The above surfactant is not commercially viable today on account of mass production problems due to the high priced coconut oil and sucrose.

Nutritional and Health Aspects of Coconut

Coconut is regarded as the symbol of health and prosperity (Thampan, 1996). For example, tender coconut is not merely a thirst quenching drink, but a health drink that has numerous medicinal properties. The use of tender nut water as an intravenous fluid for patients has been reported from many countries. Studies have shown that tender coconut water is an effective fluid for oral rehydration with its high mineral content, vitamins and minor constituents such as free amino acid and nucleic acids (Shivanandiah, 1970; Thampan, 1996).

Coconut kernel and coconut oil are considered to be health friendly foods. Coconut kernel contains protein and dietary fibre. The major protein present in the coconut kernel is globulin having high concentration of arginine, which helps to reduce the serum cholesterol. The dietary fibre present in kernel assists digestion and helps to reduce serum cholesterol. Coconut oil which is rich in medium chain fatty acids facilitates easy digestion and rapid absorption in the human body. Coconut oil is quickly turned as fuel and energy in the body. Coconut oil consumption increases the HDL, cholesterol (good cholesterol) and does not cause elevation of LDL cholesterol (bad cholesterol) (Thampan, 1994). Coconut oil has various nutritional and medicinal qualities which makes it a preferred oil in the ayurvedic preparations. It has also been reported that the lauric acid present in coconut oil keeps infants free from getting viral or bacterial infections. According to naturopathy, coconut milk with honey is a very old medicine for the cure of skin disease and stomach disorders. Fresh toddy (Neera) tapped from coconut palm is a very good health drink.

The coconut palm enjoys a pride of place in Ayurveda. The inflorescence of coconut palm has therapeutic use. It is cooling and effective against dysentery, hemophilia, diabetes and leucorrhea. It is also an ingredient in a decoction used to treat syphilis. The midrib of the leaves is used for preparing decoctions to correct and cure breathing troubles and upward movement of Vata and also to serve as a medicine for cholera. The tender roots of the palm are used in medicines to heal uterus complaints and for ascites or abdominal dropsy. As an ingredient in decoction, they are effective against fever and in paste they stimulate the discharge of urine. A decoction of the palm roots and the bark of Asoka *(Saraca indica)* is prescribed to cure infertility. Likewise, the ash of coconut fibre is astringent and used to stop bleeding. Nalikerapakam and Nalikeraksharam (ash of coconut kernel) are effective antacids widely prescribed (Warriar, 1996).

Tender coconut water is recommended in cases of gastroenteritis. It is very effective to prevent dehydration of body tissues during serious cases of diarrhoea and vomiting. It is an effective remedy against sunstroke. It moderates heart beat, strengthens heart muscles and ameliorates body temperature. Either as such or in combination with selected herbs, tender coconut water forms a household remedy against many ailments such as cholera, indigestion, constipation, piles, insomnia, hepatitis etc., For children, tender coconut water is an effective antidote against round worm induced vomiting. In cases of chicken pox and small pox, tender coconut water is specially recommended to cool the system. Tender coconut water infusions are useful in the treatment of urethritis and related diseases. Scientific preparations made of coconut, tender coconut water, or inflorances are proved high effective for curing dental disorders, diabetes, plague, excema, ulcer, diseases of eye etc., (Pillai, 1979, Pillai, 1982) *Elaneer Kuzhampu,* a lotion, prepared with coconut water is a reputed *Anjana* (collyrium) used all over India (Nandakumar, 2000).

In pediatrics, for the promotion of the health of children, from the very stage of the newborn baby, milk of the coconut kernel and various preparations of coconut oil are widely used. Coconut milk is used to clean the body of the baby from the *Verrni casseosa (Ulbam)*. The oil prepared by boiling the coconut milk is given to babies for internal use. It is very nourishing and used in place of cod liver oil. It is also applied in the eyes. For all skin diseases of children, known as *Karappan,* medicated coconut oil has the first preference. In cough and cold and for diseases of ear and secretions from mucous membranes, medicated coconut oil is effective. In many cases coconut oil is given for internal use particularly in the dried states of Kapha (phlem) to expel them (Rama Varier, 1961).

In the preparation of medicated oils, coconut oil is used.

There are many preparations with coconut oil. Many are home preparations using coconut kernel and herbs like Sarasaparilla for skin diseases, acidity conditions and for promotion of health. Cosmetics prepared in coconut oil are also traditionally used. Oil medicated with the roots of coconut palm is used to apply on head to cure troubles of the head. Coconut oil is valued for it's antibacterial and antiviral properties (Vaidyarn, 1996). Coconut oil is a very good emollient on human scalp/hair and has effects such as moisturisation glossiness and adhesion (Adhikari et al., 1996)

Other parts of the coconut palm also find useful applications. The distillate of coconut shell separated from the tar is employed to cure warts, excrescences and skin troubles including eczema (Thampan, 1996).

Post Harvest Processing, Product Diversification and Value addition in Coconut

In spite of the fact that India has made impressive strides in area, production and productivity of coconut, the processing industry did not grow to absorb the increased production which has been quite detrimental to progress of the coconut economy. Although coconut is one of the most versatile crops with tremendous economic potential it has not been fully exploited for product diversification (Aravindakshan, 1996).

Coconut industry can be grouped in to two major categories viz. conventional and non-conventional. Copra making, oil extraction from copra and manufacture of coir products are the major conventional coconut based industries in the country. Desiccated coconut powder manufacturing is also a traditional industry in the country, even though the total annual production is only around 30,000 tones per year. These traditional industries support coconut culture at large.

More than 50 percent of the coconut production is utilized as raw nuts for culinary and religious purposes, about 33 per cent is converted into milling copra which in turn is crushed to yield coconut oil and 7 per cent into edible copra. About 40 per cent of the total coconut oil produced in the country is used for cooking purposes, 50 per cent for toiletry purposes and the balance 10 per cent is utilized for industrial purposes. The remaining ten per cent of the coconut production is for manufacturing kernel based products mainly desiccated coconut, as tender nut for drinking and seed purposes (Poduval, 2002).

The price of coconut in the country moves according to the ruling market price of coconut oil. This is mainly because of the fact that post-harvest processing sector of coconut is mainly confined to extraction of coconut oil, which utilizes only 33 per cent of the total supply of coconut in the country. Moreover, price of coconut oil is subjected to seasonal variations according to the demand and supply (Singh, 1998). With the ability of cheap vegetable oils, coconut oil has also become vulnerable to substitution mainly in the household and industrial sector (Thampan, 1993). However, coconut has the advantage of multiple uses and it's dependency on coconut oil can be reduced by product diversification and by-products utilization.

Coconut industry in the country is currently undergoing structural changes particularly due to 'product diversification and value addition. Technology development for product diversification and value addition of coconut did not get its due attention until 1980s. Subsequent to the formation of the Coconut Development Board in early eighties, technological research has become a priority area and development of appropriate processing technologies was given adequate attention. As a result, various diversified coconut products entered the market, which in turn paved the way for better utilization of coconut palm parts.

During Nineties, the Coconut Development Board developed a few technologies for the production and commercialization of some of the non-conventional value added products under its sponsored programme. The technologies for preservation and packaging of tender coconut water and manufacture of coconut cream, coconut cream powder, natural vinegar from coconut water, nata-de-coco etc., have been developed. These technologies have since been transferred to some entrepreneurs for commercial production, and these products are slowly being introduced and promoted in the domestic market.

Commercial utilization of coconut in India was confined to products like copra and coconut oil. The trend has now changed due to the fluctuating market prices of these products and product diversification in coconut is gaining momentum. The technological advancement in the processing of coconut and preference for ready to use processed foods among health conscious people across the globe created worldwide market for processed coconut food products and beverages, desiccated coconut, coconut milk cream, coconut milk powder, packed tender coconut water, skim milk based beverages etc.

Constraints in the Production, Processing and Marketing of Coconuts

In spite of all the efforts in the fields of research and crop production, the results reached only a relatively small number of farmers, predominantly those living in easily accessible areas where an efficient extension service has been operating. Farmers living in remote areas where coconut plays an important role in the daily subsistence of the people did not usually benefit at all. Considerable improvement was achieved in the quality of plant material that has been available to farmers, including those in countries where research in coconut has taken place for almost a century. However, it had only little impact on the average yield level.

Although research efforts on coconut in India had enabled the evolution of high and early bearing hybrids, development of coconut-based management practices and cropping systems and control of pests and diseases, the average yield per palm had not increased over the last 20 years (Bavappa et. al., 1976). It is estimated the mean world copra yield per hectare to be lower than 600 Kg, whereas in research stations yields ten times as high had been obtained (Martin et. al., 1979). The low yield of coconut smallholdings may be attributed to a number of factors. For Malaysia, inherent low fertility of the soil or depletion of its fertility, high per cent of old palms (over 50 years of age), flooding and poor drainage, poor farming practices, infestation by pests, neglect resulting from shortage of labour and the discouragement owing to poor prices and low yielding varieties are the major factors contributing to low yield (Lebai Mat, 1979).

In Tanzania, annual fire, rapid bush regeneration, planting of food crops (without fertilizing), complex land tenure unsuitable for the development of a perennial crop, reluctance to eliminate old palms and to replant are the factors contributing to low yields of small holders (Kasembe et. al., 1968).

A survey conducted in Sri Lanka by Liyanage et. al., (1986) revealed drought, lack of funds, price instability, lack of technical know how, problems of timely availability of labour, availability of planning materials and theft as most important problems in order of importance that were faced by farmers.

In the Philippines, Arboleda et. al., (1986) reported that the main problems faced by farmers were inadequate funds, delayed release of funds, high cost and/or absence of agro-inputs, difficulty in marketing and lack of processing and storage facilities.

A meeting of UNDP/FAO working Group on Coconut-Based Farming Systems (CBFS) in Thailand (Anonymous, 1989) identified marketing, credit, information, inputs, farm size, land tenure, farmers education, storage, transportation etc., as the factors influencing CBFS. In India, the factors identified as limiting the production and productivity are low productive capacity and variability of the local coconut cultivars and favourable climatic factors especially rainfall pattern, inadequate irrigation facilities, prevalence of pest and diseases in major coconut growing tracts, use of unsuitable land for coconut cultivation, large number of small sized holdings, non-adoption of improved farming practices, non-availability of superior planting materials, and escalating cost of production

The major factors hampering the coconut industry especially the processing and marketing sectors are labour intensive operations, post harvesting losses, lack of adequate research and development efforts, insufficient consumer awareness, lack of quality standards, violent market fluctuations, absence of cost effective packaging, debt bondage resulting in forced sale of coconuts, absence of bargaining and dearth of organisational support and facilities, increasing competition from low priced products, varied tax structure, and negligible exports

SUM UP

In conclusion the Indian Coconut cultivation poses a very bright future despite the constraints and problems beset with the industry in certain areas. Available infrastructure, trained manpower and wide range of climatic conditions in the country are indicative of best capabilities. At the same time, these resources have to be effectively utilized to harness the best for making coconut industry more competitive and dynamic. India will continue to be the premier producer in the world and also assume leadership in the trade.

REFERENCES

1. Adhikari, S., and others, Coconut Oil as a Hair Oil, National Meet on Coconut Oil, Coconut Development Board, Kochi,1996.
2. Ajith Kumar, P, Coir Industry in India, *Processing and Marketing of Coconuts in India,* CDB, Kochi, 1994.
3. Aravindkshan, M., Emerging Trends in Coconut Culture and Challenges to Research, *Indian Coconut Journal*, CDB, Kochi, (Vol. 26, No. 9), 1995.

4. Aravindakshan, M., Challenges to the Coconut Industry in India and Strategy for making it competitive, Coconut Development Board, Kochi, 1996.
5. Aravindkshan, M., India Attains the Premier Position in Coconut Production, *Indian Coconut Journal*, CDB, Kochi, (Vol. 28, No. 3), 1997.
6. Arboleda, and others, Coconut —problems and approaches. The Philippine experience, International Consultation on Plantation Crop Diversification. FAO (Italy), November 1986.
7. Bavappa, K.V.A., and others., Coconut and Arecanut based Cropping Systems, J.Plantain. Crops, Kochi, Volume 14(2).
8. Bernd Kretschemer and Amal Kumar Bose, (1999). Protein shifting in Coconut processing. *Indian Coconut Journal,* 30, pp. 36-37.
9. Boceta Norberto, M. The Challenges of making the coconut industry globally competitive, Coconut Development Board, Kochi, 2000.
10. CFTRI, 1999. Home Scale Processing and Preservation of Fruits and Vegetables. CFTRI, Mysore, pp. 11-12.
11. Coconut Development Board, Proceedings of the National Seminar Report on Processing and Marketing of Coconuts (SPAMCO) held on 13th January, Bangalore, 1994.
12. Coconut Development Board, Integrated disease management in coconut, *Indian Coconut Journal*. Vol. XXIX, No. 4, Auguest 1999.
13. Coconut Development Board, Tender Coconut, Kochi, 2001.
14. Chettiar Lakshmana, *Folklore of Tamil Nadu*, National Book Trust, New Delhi, 1973.
15. Child, R. *New Biology, The Coconut*, Longman Group Ltd., London, 1953.
16. Coconut Development Board. (2001). *Tender Coconut,* Coconut Development Board, Kochin.
17. Coconut Development Board, Demand and supply of coconut products and consumers buying behaviour in Tamil Nadu; Research Report, 2003.
18. CPCRI. Annual Report, Central Plantation Crops Research Institute, Kasaragode, Kerala, 2000.
19. Dwarakanath, Vinegar Fermentation with Special Emphasis on Possibilities of Utilisation of Matured Coconut Water, *Processing and Marketing of Coconuts in India*, CDB, Kochi, 1994, pp. 44-51.
20. Enig Mary. G., Rationale for use of lauric oils as physically functional foods for HIV infected individuals, *Indian Coconut Journal*, September, 1999.

21. Gnanasekar Francis., Globalisation and WTO – Indian, Insight, *Kisan World,* March, 2003.
22. Jon J. Kabara. Health oils from the tree of life-nutritional and health aspects of coconut oil, *Indian Coconut Jouanal,* No. 8, December, 2000.
23. Kadalli and Suseela Nair., Manurial value and efficiency of coir dust based enriched super compost, *Indian Coconut Journal.* Vol.XXXI, No.3, July, 2000.
24. Kasembe, J.N.R. and Currah, I.A. Coconut experiments in Chambezi, Tanzania, Paper, FAO Technical working party on Coconut Production, Protection and Processing, Jakarta, Indonesia, 1968.
25. Katar Singh., Cooperatives – Challenges in emerging economy, NCDC Bulletin, December, 2002.
26. Kaul, G.L., Product Diversification in Coconut —World Scenario (Processing and Marketing of Coconuts in India), CDB, Kochi, 1994.
27. Kaushal H,. Optimum Utilisation of Coconut, *Processing and Marketing of Coconuts in India,* CDB, Kochi, 1994.
28. Lalithamma, K. (1996). *Pharmacopeia.* Government Ayurvedic College, Thiruvananthapuram.
29. Liyanage, D.V., and others., Coconut Breeding in Indonesia, *Cocos* Indonesia, 1986, p. 4.
30. Madhavan, K., and other., *Recent Developments in the Harvest and Post-harvest Technology of Coconut Breeding and Management,* Trissur, 1991.
31. Mall, P.P., Desiccated Coconut Industry in India—Problems in Internal and Export Trade, *Processing and Marketing of Coconuts in India,* CDB, Kochi, 1994.
32. Markose, V.T., *Domestic Marketing of Coconut Products in India,* APCC, Jakarta, Indonesia, 1992.
33. Markose, V.T., Coconut Processing Sector in India—An Overall Perspective, *Processing and Marketing of Coconuts in India,* CDB, Kochi, 1994.
34. Markose, V.T., Fifty years of Coconut Development in India, *Indian Coconut,* June 1998, p. 29.
35. Mathew A.C., Energy Efficient Irrigation System Design, *Kisan World,* June, 2003.
36. Menon, K.P.V and others, The Coconut Palm—A Monograph, Ernakulam, Indian Central Coconut Committee, 1958.
37. Muralidharan, K. and others., Coconut Development in India—The Past, Present and Future, *Indian Coconut Journal,* 1998, p. 28.

38. Nair, M.K., and others, Coconut Research in India, *Coconut for Prosperity*, 1996.
39. Nampoorhiri K.U.K. Milestones in Coconut Research, *Indian Coconut Journal*. May, 1998.
40. Namboodhiri, K.U.K and Singh, H.P., *Trends in Coconut Research and Development in India*, CDB, Kochi, 2000.
41. Namboodiri, K.R.R. and Nair, G.S. (1998). Nattarivukal. *Karshakashree*, August, 1998, pp. 34-35.
42. Nandakumar, T.B. (1997). Thenginkallu. *Indian Nalikera Journal*, April, 1996. pp. 10-11.
43. Nandakumar, T.B., Coconut palm products in social and religious functions in India, *Coconuts Today*, 1985 P. 3.
44. Nandakumar T.B., Pleasant Fluid of Coconut, *International Coconut Conference*, Coconut Development Board, Kochi, 2000.
45. Norberto M. Boceta., The Challenges of Making the Coconut Industry Globally Competitive, COCHTECH seminar, CDB, Kochi, 2000.
46. Panicker, T.R.R.K. (1960). Brihath Asavarishta Manjula. Shree Lakshmivilasam Book Depot, Ernakularm.
47. Poduval Sreekumar., Diversification and Value addition in Coconut Processing, *Sustainable Production and Utilisation of Coconut*, CDB, Kochi, 2002.
48. Punchihewa P.G., Current Status of the Coconut Industry, 37th COCOTECH International Coconut Conference, Chennai, July 2000.
49. Ratnambal. M.J. (1999). Varieties suitable for Tender Coconut, *Indian Coconut Journal*, 30, 64-66p.
50. Ratnambal, M.J. and others, Floral Biology of some coconut accessions. *Journal of Plantation Crops*. 2003, p. 31.
51. Rethinam, P., Management of Coconut Palms, *Coconut Research and Development*, KAU, Vellanikkara, 1991.
52. Rethinam, P. and Nandakumar, T.B. Coconut Statistics (Supplement), CDB, Kochi, 2001.
53. Rethinam, P., Coconut Industry in India, CDB, Kochi, 2001.
54. Satyavati Krishnankutty, Products from Coconut Water, *Processing and Marketing of Coconuts in India*, CDB, Kochi, 1994.
55. Singh, H.P., Coconut Industry in India—Challenges and Opportunities, CDB, Kochi, 1998.
56. Singh, H.P. and Mathew, M.T., Sustainable Production and Utilisation of Coconut, CDB, Kochi, 2002.
57. Singh, H.P., Strategic Agenda for Improving Competitiveness of Coconut Industry in the Midst of Challenges from Competing Products paper presented in XXXIX Cocotech, Thailand, 2002.

58. Sreekumar Poduval., Utilisation of Coconut by-Products in India, *Processing and Marketing of Coconuts in India*, CDB, Kochi, 1994.
59. Sreekumar Poduval, Ajay. M. Pillai and Balachandran Nair, (1998). In : Harvest and Post Harvest Technology of Plantation Crops. Technologies for product diversification and by product utilization in coconut, p. 70.
60. Subburaj, B., Coconut Processing industry in Tamil Nadu, with special reference to Kangayam, paper presented in the National meet on Coconut, held at Kochi, November, 1996.
61. Subburaj, B., Coconut Oil Mills in Kangayam Region, Erode District, Tamil Nadu. *Indian Coconut Journal*, 1997, p. 28.
62. Subburaj, B., Participatory Appraisal on Marketing of Coconuts—Practices, Problems and Prospects in Tamil Nadu, GRI, Gandhigram, 1998.
63. Subburaj, B., Coconut Marketing in Tamil Nadu, Ministry of Rural Areas and Employment, Department of Rural Development, GOI, New Delhi, 2000.
64. Subburaj, B., Coconut Production, Practices and Problems in Tamil Nadu, A Research Monograph, Gandhigram Rural Institute, Gandhigram, 2004.
65. Subburaj, B., Coconuts in Regulated Markets. Research Monograph, Gandhigram Rural Institute, Gandhigram, 2004.
66. Thampan, P.K., Glimpses of Coconut Industry in India, CDB, Kochi, 1988.
67. Thampan P. K. (1996). Coconut for prosperity. Peekay Tree Crops Development Foundation, Kochi, Kerala, pp. 247.
68. Thampan, P.K., The Coconut Profile of India, *Glimpses of Coconut Industry in India*, CDB, Kochi, 1998.
69. Venketachalam, V. and Thomas, P.T., Investment opportunities in Coconut in Tamil Nadu, *Indian Coconut Journal*, June 1997.
70. Nandakumar, T.B., Thenginkallu. *Indian Nalikera Journal*, April 1996. pp. 10-11.
71. Bernd Kretschemer and Amal Kumar Bose, Protein shifting in Coconut Processing. *Indian Coconut Journal*, 1999, pp. 30, 36-37.
72. CFTRI, Home Scale Processing and Preservation of Fruits and Vegetables. CFTRI, Mysore, 1999, pp. 11-12.
73. Ratnambal. M.J., Varieties suitable for Tender Coconut, *Indian Coconut Journal*, 1999, pp. 30, 64-66.
74. Thampan P. K., Coconut for prosperity. Peekay Tree Crops Development Foundation, Kochi, Kerala, 1996, p. 247.

4

Production Practices of Coconut Growers

Introduction

Coconut is a perennial palm alive for 80 to 100 years with an economic life of 60 years or more depending upon local conditions. The palm takes about 7 to 10 years to commence fruiting and another five years or more to come to the stage of full bearing. Thus, only after 15 or 20 years will the grower is in a position to reap the reward for his labours. If the original planting material used happens to be poor in quality, it will result in the establishment of a plantation giving poor yields and turn out to be a source of loss to the grower as long as the plantation lasts. There is nothing that he can do then to remedy matters except to cut and remove the unproductive and uneconomic trees. These facts will underline the need to plant quality planting material that will ultimately give good yields. The coconut palm is a monocot and is capable of being propagated only through seed. As the palm is heterogeneous and cross pollinated in nature, the progeny is likely to be variable, but certain precautions, if observed, are found to help in the production of quality planting material. As a result of intensive investigations done in the important coconut growing countries of the world, it has been possible to fix criteria and to standardise nursery technique by which a large percentage of quality seedlings can be produced at relatively cheap cost.

Production of Planting Materials

In order to obtain higher yields from plantation, utmost care is required right from the stage of selection of quality planting materials as coconut cultivation involves substantial investments.

Climate and Soil

The coconut palm is found to grow under varying climatic and soil conditions. It is essentially a tropical plant growing mostly between 20°N 20°S latitudes. However, a rainfall of about 2000 mm per year, well distributed throughout, is ideal for proper growth and maximum production.

Coconut is grown under different soil types such as loamy, laterite, coastal sandy, alluvial, clayey and reclaimed soils of the marshy low lands. The ideal soil conditions for better growth and performance of the palm are proper drainage, good water-holding capacity, presence of water table within 3m and absence of rock or any hard substratum within 2m of the surface.

Selection of seed garden: The coconut garden from which the seed nuts are collected for propagation should be ideally located in pest and disease free areas. The garden should have a good number of heavy bearers and should give consistently good yield with good quality nuts. The palms in the garden should ideally of the same age group, growing under identical conditions.

Selection of mother palms: Selection of mother palms of desirable traits has been found to increase the efficiency of nut production by 15 to 50 per cent. Mother palms are selected based on the visual phenotypic traits and easily discernible characters. Ideal mother palms will have stout straight trunk, short strong bunch stalks with closely spaced leaf scars and a full crown with 30-40 average sized fronds. There will be short fronds oriented spherically or semi spherically on the crown and a large number of inflorescence (12-15 per year) evenly distributed round the crown one in each axil.

Generally, the mother palms selection is restricted to 10 per cent of the total palms in each garden. Such selected palm should be yielding about 100 nuts per palm per year. The weight of husked nuts will be 650-700 gm. and copra weight 150-200 gm. Yield can be roughly estimated by counting the number of nuts on the crown from oldest to youngest. Selected palm having a large number of spikelets with 2-3 female flowers on these are found to possess high per cent of fruit set and gives stable yields. Palms having short pre-bearing stage and higher leaf production capacity are known to be early and higher yields. Even young palm which show early regular bearing habit and other desirable traits may be selected as mother palms during 3rd or 4th year of bearing.

Collection and selection of seed nuts: 11-12 month old fully ripe nuts from middle and bottom portions of heavy bunches germinate better. The maturity can be assessed by tapping the nuts which produces a resonant ringing sound. On chipping a portion of the husk a distinct browning of inner fibrous tissue will be seen in matured nuts. Such seed nuts should have a small quantity of water. Spherical shaped nuts are found to germinate early and give more vigorous seeding. The selected seed nuts should be weighing 1.8 - 2.4 kg with a copra content of about 160-200 gm. Such heavy seed nuts will float vertically if put in water with its stalk end up. Thin husked nuts are likely to give more vigorous seedlings. Light undersized, cracked, un-uniform nuts should be discarded while selecting the nuts. Seed nuts are collected during January - April in the west coast and February - June in the east coast as the nuts are big in size with maximum copra content. They germinate early and give out vigorous seedlings.

Storage of seed nuts: In order to preserve the seed nuts without desiccation, it is necessary to store the selected nuts under shade for almost a month before sowing in the nursery, during which time the husk becomes dry. If longer storage is required the dried nuts are placed in a thick layer of sand with stalk end up and completely covered with sand up to a thickness of 5-7 cm. Additional layers are built up over the first layer depending on the number of nuts. Five layers can be accommodated thus. This procedure will help the seed nuts to remain viable for a period of 8 months or more.

Seed treatment: Soaking the seed nuts in water for about 2 weeks is found to promote quicker germination especially when dry conditions prevail after sowing the nuts in nursery.

Nursery practices

Raising the seedlings in a nursery bed before planting out in the main field is essential for selection of the seedlings, for better upkeep of the seedlings, and economical management of inputs like water, nutrients etc.

Placing the seed nuts in a germination bed before planting in the nursery could also be adopted to accommodate about 10 times more seed nuts. Selection of seedlings based on early sprouting could be made in the germination bed so as to reduce the number of nuts to be sown in nursery. The germination bed should have friable well drained soil. Sandy soil helps in reducing the termite attack. Sprouted seed nuts could be taken out for planting in nursery bed when the sprouts reach a height of 15-20 cm.

Nursery site: The site selected for nursery should be open and have light, friable well drained soil. Availability of water is extremely important to facilitate satisfactory irrigation of the nursery. Sandy soils are preferred for nursery beds as it well help in reducing termite attack, facilitate drainage and helps in easy lifting of seedlings. If the soil is not sandy, it is better to remove the soil to a depth of two feet and fill up with sand. Nursery should not be raised under dense shade as the seedlings tend to be lanky. Provisions of raised beds are required in places where the drainage is blocked during heavy rains. However it may be inconvenient in areas where flood irrigation is adopted. In such cases, channels should be opened up to drain out the excess water. Treating the nursery bed with Phorate @ 10 grams/ square metre or Chlorphyriphos @ 0.05 per cent concentration is beneficial against white ants and grubs.

Spacing in the Nursery: To facilitate manual watering, weeding etc., the width of the nursery bed is to be adjusted to accommodate 4 to 5 rows of seed nuts (2m. width) and convenient lengths (3m.). A spacing of 30 cm. (measured from centre of the nut) from nut to nut and 40 cm. between rows and 80 cm. between beds is optimum. Thus 6,000 nuts can be accommodated in 0.1 ha. If seedlings are to be retained in the nursery for more than 18 months it is better to provide wider spacing of 90 cm. between nuts.

Time of sowing seed nuts: The appropriate time for sowing seed nuts may vary from tract to tract depending upon monsoon sowing the nuts at the commencement of monsoons will provide optimum conditions for germination and subsequent growth. Before sowing the seed nuts, it is necessary to examine each nut and discard those in which water is completely dried up.

Method of sowing the seed nuts: The nuts are buried in trenches 20-25 centimetre deep in seed bed in a horizontal or vertical position keeping the stalk up just above the soil surface. Horizontal sowing with the widest of the three segments placed upper most has been found to give higher per cent of germination. Such seed nuts sprout early and give out vigorous seedlings with thicker girth at collar end because the nut water remains close to the embryo. Vertical planting of seed nuts in nursery bed is advantages as it is easier to handle the seedlings and can be compactly packed especially when the seedlings are to be transported to a longer distance. Moreover as the un- germinated seed nuts are removed after 5 months from the nursery. Higher germination value in horizontal sowing has no significance.

Care and Management of the nursery: Management of the nursery involves regular irrigation, weeding and mulching, providing shade and control of pests and disease so as to have early and maximum germination and subsequent healthy seedlings. Watering is indispensable in the nursery bed. Irrigation depends upon the type of soil, distribution of rain fall, temperature and humidity. In general watering every alternate day @ 10 litres per square metre per day is necessary if the nursery is raised in sandy soil with little shade or twice a week if adequate shade is provided. Irrigation could be skipped during the days following rain fall of more than 10 mm. Sprinklers or perfo sprayers are ideal for coconut nurseries. Small nurseries could be watered by hoses or pots.

Organic mulching will help to keep the soil moist and reduces the weed growth, in addition to promoting early and better germination. Periodical weeding is required for the proper growth of seedlings. Adequate plant protection measures have to be adopted in the case of pest and disease incidence. Treating the soil with Chlorpyriphos (50 per cent WP) and spraying the seedlings once a week with Dithane M 45 (2 gm/lit. of water) would be beneficial against pests and fungal diseases. Coconut seed nuts contain ample food reserves for them to sustain germination and hence manuring is not generally recommended for nurseries. However application of Potash @ 60 gm../seedling after the initial 2-3 roots emerge (14 weeks) after sowing may be useful to maintain seedlings vigour.

Emergence of sprouts and growth of seedlings: The seedlings begin to germinate in 8-10 weeks after sowing and in 5 months, 85-90 per cent of the seed nuts would have germinated. First leaf appears in the second month of germination and the root emerges one month afterwards. By sixth month, the entire cavity of the nut will be filled with the apple or haustorium. Those nuts which have not germinated even after 5 months and the dead sprouts should be removed from the nursery and may be converted to copra.

Selection of seedlings: Just like the mother palm and seed nut selection, the selection of seedlings alone is reported to increase the yield by 10 percent. Selection is based on the age of the seedling, early germination and vigour of the seedling. In general the younger seedlings of 7-9 months age have better chances of early establishment in the main field. At this stage, 40-50 per cent of the kernel in the nut will be intact and the seedling can utilize the food material during the initial period of transplanting. In the case of aged seedlings with lesser or no food reserves, the transplantation shock and root injury will be considerable and will lead to delayed establishment in the field due to its sole reliance on soil for nutrition. When such aged seedlings are transplanted to the main field they have to be provided with supports or props till they establish proper root system.

Good seedlings will have a healthy appearance, with six leaves after one year of sowing, about 10 cm girths at collar end and show early splitting of leaves. Early germination of seed nuts is associated with faster rate of leaf productions in the seedlings. Production of more number of leaves and early splitting of leaflets in the nursery is again correlated with early flowering and early bearing character. If the above standards like early germination, number of leaves, early splitting of leaflets, girth at collar etc., are adopted for the selection, it is possible to get 60-65 seedling for every 100 nuts sown. Seedlings without leaflets or with very small leaves, erect and narrow leaves, widely spaced leaflets, elongated petioles, narrow girth at collar and short dump seedlings should be invariably discarded.

Removal of seedlings from nursery: The seedlings should be lifted from nursery gently by spade after loosening the soil around the nuts and should not be pulled by force. The seedlings than should be transplanted as early as possible in the main field. Under favourable weather conditions, transplanting should be done preferably within 10 days of removal from nursery, even though the seedlings could withstand up to 1 month with adequate watering and shade. Under favourable condition the seedlings would establish after 4 weeks of their- removal from nursery.

Planting coconuts in polybags: The planting of seed nuts in poly bags for raising seedlings is a method developed by the IRHO in Ivory Coast in1969. The advantages are: a) better seedling development with regular fertilizer dressing b) safer handling of seedlings c) no root damage while transplanting d) better rooting in the field e) no shock during transplanting and f) uniformity in stand even after gap filling.

Recommended poly bag types: Black polyethylene (polythene) bags of 40 cm. length and 40cm. height and 0.18 to

0.20 mm. are used for raising seedlings. These bags are to be provided with holes of 5 mm. diameter in the lower part of the poly bag. The spacing for the holes may be 5 cm. between holes and 7 cm. between rows of holes. Nearly 16 poly bags of the above size will weigh 0.900 gm. – 1kg. The bags are filled up to its two third with fertile topsoil or with a mixture of soil and manure or compost. The bags are placed in a levelled bed at a spacing of 60 cm. The germinated nut with sprouted portion on the same side is placed in each bag and the bag is filled with soil up to the collar end. The top of the seed nut should be 2 cm. above soil level and the soil level should be 4 cm. below the rim of the poly bag. The seedling have to be watered on alternate days @ 10-15 mm./day. The seedlings are kept in the nursery till they reach about 4 feet from nut to unfurled leaf. Nearly 18,000-20,000 seedlings can be accommodated in a triangular fashion with a spacing of 80x80x80 cm. or 60x60x60 cm. per ha. When the seedlings are transplanted

Garden Establishment

Selection of land: The area selected for coconut new planting should have suitable climate to ensure proper growth and productivity of the palm. The soil profile of the area should be suitable for coconut culture especially with respect to type of soil, thickness, texture, depth and nature of subsoil etc. Preparation of the land for planting will depend on the topography, ground water level and soil type of the area. If the area is undulating or sloppy, soil conservation measures may have to be adopted. If the ground water level is high and the area is subjected to water logging, preparation of raised mounds may be necessary. It may also be necessary to tap the ground water potential by digging well, bore wells or tanks for irrigation. In brackish water area, raised bunds of 5-8 m. in width are to be built up by dumping clay and sand in alternate layers.

Clearing of new land: Clearing of new land for coconut cultivation is more or less the same as for any other crops. Filling, de stumping and burning of the shrubs may be required in areas where the vegetation consists of trees and shrubs. Filling is generally done one and half months prior to planting during dry season. All the shrubs logs and stumps are to be burnt thoroughly as otherwise they get rotten and become breeding sites for beetles. Vegetation consisting of grasses and small bushes may be cut and removed by hand or by a tractor or bulldozer depending on their severity. In cases where hard grasses exist, repeated weeding may be required to exhaust the grass regrowing from underground stolons. Ploughing at 8-10 days intervals may be useful in such cases.

Layout of the plantations: The layout of the plantation has to be planned before clearing and depends on the size of plantations. The layout consists of division of the land in to convenient blocks, laying out roads, drainage channels, lining, putting the planting holes, etc., on flat lands blocks of equal size are made with roads dividing each block. Secondary roads within each block could be equal in size to the space between two rows of palm. On sloppy lands, roads should be laid out on the contour or take a gradual descending route so as to prevent erosion. In such areas, soil conservation measures like, contours bunding, terracing etc., may have to be carried out depending on the slope. Coconuts are not grown in areas where the gradient is more than 20 per cent.

Drainage is necessary for soil erosion especially in areas where water table is high. Drainage system depends on local climate, soil type and topographical conditions. Drainage system and road system should be compatible. On either side of the road drainage is inserted to prevent logging of water on the road. Lining should be carried out before laying out road and drains. It starts from a base line along east-west with a length of about 10 planting distances. On both ends other lines will be set out at 90° to the first line in the case of square system of planting or at an angle of 60° in the case of triangular system of planting. Between the ends of the perpendicular line, a line parallel to base line is drawn and the whole would form a diagonal or rectangle which forms the basis for the remaining lining system. Stakes of about 1.5 – 2 m. length may be used to mark the planting sites. A wooden frame (100 cm. square) may be used to mark the size of the pits for planting.

Spacing and planting density: Spacing would depend upon the water availability, type of soil, variety or whether the crop is grown as mono crop or in association with other seasonal or perennial crops. In monoculture, the spacing provided should be such that the leaves of adjacent palms should not over lap when fully grown, so that the palm will receive maximum sunshine for its growth without undue competition. The spacing generally varies from 7.5 m. in a triangular system (210 palms/ha.) to 10 m. in a square system (100 palms/ha.). The recommendation is to have a spacing (247 plants/ha.) in laterite gravely soils where coconut is grown under rain fed conditions.

In general, for a spacing of 7.5 to 9 m. between plants will be useful for a mono crop on triangular system or for a multi-crop on the square system in red sandy loams and coastal sandy belts. In the case of dwarfs, a spacing of 6-6.5 m. on the triangle with a north south orientation of rows is recommended under normal conditions. In the case of single hedge planting a distance of 9 m. between rows and 5 m. in the rows is optimum. For double hedge planting, 5 x 5 m. in rows and 9 m. between pairs of rows would be ideal. When planting is done on the sides of channels or on bunds in single rows, a closer spacing of 6 m. between plants will be adequate.

Systems of planting: Among the different methods of planting adopted, the square and rectangular systems are most common. When grown as a mono crop, the triangular method is beneficial as it will accommodate 15 per cent more palms than square method.

Table: 4.1

Number of Palms per ha. under Different Methods of Planting

Planting distance (m.)	*Number of palms/ha.*	
	Square Method	*Triangular method*
6.0	278	320
6.5	237	273
7.0	204	236
7.5	178	205
8.0	156	180
8.5	138	160
9.0	123	143

In the case of hedge system, the number of palms could be increased by 30 per cent so as to facilitate removal of late flowering and low yielding palms during the 3rd to 7th year of planting. By tenth year, a very uniform stand of good yielding palm would be obtained without affecting the required density. This system is useful when intercrops are grown. The spacing adopted in this method is 5-6 m. within rows and 9 m. between rows accommodating about 185 palms/ha.

Another system of planting is double hedge row method, which facilitates lesser shade in the early years of multiple cropping. When annual food crops are grown as intercrops, this method is beneficial as the inter cropping could be practiced for longer periods.

In the case of double hedge system a spacing of 6 m. x 6 m. within rows and 9 m. between pairs of rows could be adopted with a north to south orientation of rows for maximum availability of solar radiation. Under this system, about 217 plants could be accommodated in an hectare.

Digging pits for planting: Planting holes or pits are made after peg marking at appropriate distances, usually 2-3 months prior to planting in order to improve the surrounding soil through exposure to air. The size of the pit depends upon soil texture, nature of land and ground water level. In general, light soils require smaller pits than heavy soils. In hard laterite soils, a pit of 1.3 m. cube is required. In the east coast and in loamy soils and sandy loams, pit size could be 0.60 cm. In loose sandy soils, tapering pits with 1.25 m. top and 60 cm. bottoms may be adopted. In light soils, where water table goes down very deep in summer, it is advantageous to have 1 m. cube pits. In areas where water table remains near the surface or in low lying areas subjected to floods, it is advisable to plant the seedlings on raised mounds or bunds. Intervening spaces between adjacent mounds may be filled up over a period to form a continuous bund. Surface planting is not advisable as it tends to produce roots from the stem and lack in well developed root system. The surface root also dries up during summer. The surface planted seedlings are more prone to cyclone damages and in general they exhibit poor growth and gets affected by drought. Hence planting at 60-75 cm. below the surface is essential to give better anchorage and to increase the stem area from which roots emerge.

Dressing up the pits: After the pits are dug, organic refuses are decayed and burnt in the pits and sides of the pits are charred to prevent termite attack. In laterite soils application of 2-3 kg common salt per pit will soften the laterite and improve the soil structure. To avoid surface water flowing into the pit a small bund may be raised around each pit using the subsoil. Before planting, the bottom of the pits has to be loosened and filled with a mixture of topsoil, compost, or wood ash and river sand up to 60 cm. below the surface.

Addition of 5 kg neem cake per pit is advantageous to prevent white ant attack and to promote early growth of seedling. Similarly mixing of 1 kg. bone meal or rock phosphate in the filling mixture is effective for better root development and establishment of seedlings. In dry areas and in soils with low water holding capacity, it is advisable to place a few layers of coconut husks with its concave side up at the bottom of the pit. Such a practice would improve soil moisture as an absorption mass for leaching fertilizers.

Time of planting: Planting season depends upon the climatic condition. The best time would be just before the onset of monsoon. If proper shading and adequate watering is done planting could be taken up before the advent of monsoons also. In areas where water logging is possible, it is better to plant either after the heavy showers or long before the monsoons set in.

Planting: While planting, the seedling is placed about 75 cm. below the surface of the pit, on small holes made in soil. Compost-sand-mixture is placed inside the pits. After planting, the husk of the seed nut should be just visible outside. The pits are then covered fully. Care should be taken not to cover the collar of the seedling or the soil entering the leaf axils as these may cause rotting of the stem. Props or stakes may be provided for about a year so that growth and development of roots will not be adversely affected by wind or other interferences.

Planting of seedlings from polybags: The day before planting, seedlings in polybags has to be watered to keep the soil firm. Handling the seedling by the collar is to be avoided. Before placing the polybag in the pit, the bottom of the bag is to be slit of open with knife. The bags are to be removed only after planting the seedling with the bag and partly filling with soil. About 150 seedlings could be planted per man per day.

Cultural practices in young plantation: The entire exercise and investment in the selection of site, mother palm, seedling and proper care in nurseries will be of little use if not followed up by adequate care and attention of the seedling during early stages of growth upto 3-4 years of planting in main field. For the protection of young seedlings, fencing each palm with bamboo thorns or such other caging materials is advantageous until they attain sufficient heights, so as to prevent cattle grazing. During the initial stages (one to two years) and especially in areas where the summer is severe, plaited coconut or Palmyra leaves could be used for shading the seedlings until they get fully established. It is helpful in reducing the causalities and the frequency of watering.

In order to avoid stagnation of water in the pits, bunds should be made around pits and rain water if any collected in the pits should be drained off by clearing the water from pits as often as possible. Drainage has to be provided wherever necessary especially during rainy seasons. During dry weather, the young seedlings should be irrigated; this is important until the seedlings strike roots. The established seedlings up to 4-5 years of age should be watered once in 4 days @ 45 litres/seedling for its fast growth and good vigour. A mud pot containing three to four holes and having a capacity of 5 liters water placed near each seedling and replenished at weekly intervals will ensure sufficient soil moisture for the growth of the seedlings. In order to keep the soil moisture status and to reduce weed growth, it is advisable to mulch the seedling with organic matter especially during post monsoon periods and before prevailing of dry weather. It is also advantageous to plough the soil to keep the surface soil mulched and loosened.

Weeding of the area around the young palms is very important. Grasses and other obnoxious weeds around the seedlings should be kept down and mulched with organic mulches or dried weeds 1 to 1.5 m. around the seedlings. Weeds or grasses grown during rainy season should be turned down by ploughing, digging or forking at periodic intervals. Regular gap filling with healthy and quality seedlings should be done promptly every year to make good the number of weak and unhealthy seedlings removed during the 3–5 years of transplantations period. Nearly 18–20 gaps may have to be filled in second year and 7–10 in third year.

Intercrops like banana, pineapple, tapioca, yarns, paddy and millets can be grown in the interspaces during the first 5 years to augment the income. However each intercrop should receive irrigation and fertilizers treatments, without which the whole exercise may become a losing proposition. For the control of termites, leaf eating caterpillar and beetles proper control measures with insecticides should be adopted. Copper pesticides may be sprayed against budrot and shoot rot diseases.

Irrigations in coconut gardens: Most of the coconut growing tracts are pronounced with a marked dry season. Even though the coconut palm may survive considerable water stress, the yields are severely affected. Coconut palm requires adequate soil moisture at 1m depth during establishment period of 3–5 years for their normal growth and subsequent nut production. Long dry periods are harmful for coconut both to vegetative growth and nut production. The coconut palm has a spreading root system. The greater concentration of roots is in the first one meter of the soil and within a radius of 2 m. Each root has a growing point at its extreme trip. Just behind it is the whitish region about 2m. to 5m in length which absorbs water. When soils dry up for prolonged periods, the water absorbing region develops a thick wall through which water will not enter and thus palm will not be able to absorb water. It is estimated that about 28–74 litres of water is lost daily from a mature palm.

Drought symptoms appear in seedlings then first in young palms and eventually in adult palms. Drought affects the activity of growing points, reduces the production of leaves inflorescence and nut production. In older palm, premature nut fall and abscission of older leaves are the first symptoms of water stress. A long dry spell will result in a reduction of not only the number of nuts but also their size. Hence it is necessary to irrigate coconut garden at proper intervals, certainly during dry seasons. The economy of irrigation in coconut gardens depends on the sources of water, availability of water, irrigation systems adopted and cost of pump sets, fuel, etc., and also the yielding potential of palm. High yielding varieties and hybrids require regular irrigation and manuring to express their yielding potential. Similarly when multiple cropping is practiced in coconut garden the cumulative aspect of water on coconut and the intercrops may improve the economy of the system.

It has been found that when basin irrigation was given once in 7 days for west coast tall palms growing in sandy loam soils, the female flower production increased by 28.8 per cent, fruit set by 39.8 per cent and yield by 43–85 per cent. In littoral sand basin, irrigation one in 5 days resulted in 90 per cent increase in flower production and 51.4 per cent in yield. In adult palm, the first response to irrigation is the reduced premature nut fall and direct yield increase within one year. When irrigation is continued every dry season, these palms continue to give increased yield resulting from formation of more spadices and flower primordial during dry season also. Irrigation is found to increase the nut yield weight of kernel, its thickness and the total nut weight.

In order to derive the benefits of manuring, irrigation is absolutely necessary. It has been observed that high doses of fertilizer with irrigation. yielded an average of 114.5 nuts/palm as compared to 28.4 nuts on palms which received same dose of fertilizer without irrigation.

Systems of irrigations: The design, the equipment and the technique of replenishing the soil water deficit by applying and irrigation water is called irrigation system. The objective of the adopted system should be that of a uniform distribution of water in the root zone and a higher efficiency of water application.

The commonly adopted irrigation systems are surface irrigation system, sprinkler system and drip irrigation systems.

In the surface irrigation system, the common methods are flooding, basin irrigation and furrow irrigations. In slopes, this method could be adopted after terracing. To economize water, basin method is more suitable than flooding and furrow method. The basins may be of 1.75 m. radius and irrigation may be done at interval @ 1000 lit./day at 10 cm. depth once in 5-7 days.

In the sprinkler irrigation system, water pressure is applied to the surface in the form of a thin spray from above. In this system, water can be applied at controlled rate and uniform distribution and high efficiency can be ensured. It is especially useful in shallow sandy soils of uneven topography where leveling is not practicable and in areas where water or labourers is scarce. The sprinkler systems in general consists of a pump to lift water under pressure, pipes for conveying the water, sprinkler heads or nozzle and the risers which connect the sprinkler heads with the pipe line. Based on the equipment with which spraying is done, the sprinkle system is classified as rotating head type or pipe type.

By adopting sprinkle perforated system, wastage of water through surface run off and leaching of soil nutrients etc., could be minimized. With sprinklers, a given quantity of water can irrigate 2–3 times the area that is irrigated by surface irrigation.

Drip irrigation: Drip or trickle irrigation involves the slow application of water drop by drop to the root zone. This concept takes its roots in Israel during sixties. In this system, filtered water or even nutrients is applied through droppers at a controlled rate directly to the root zone. The wetting area and rate of water applied is controlled so as to optimize soil moisture status.

Advantages of Drip Irrigation

- *Saving of water:* Since water is dropped at the root zone only in controlled quantities percolation losses and surface evaporation is eliminated and the savings in water is as much as 60–85 percent.
- *Higher yield:* Since the water is applied in optimum quantities, it enhances plant growth and early flowing. Soil nutrients losses by leaching are also minimized. The yield increase in crops by drip system may be as high as 55 to 120 per cent.
- *Savings on fertilizers and nutrients:* Nearly 60 per cent higher efficiency is reported when fertilizers are applied through drip system along with water as there is no runoff or leaching and sedimentation of soils.

- *Savings on labour:* By eliminating weeding and reducing inter cultural operations, considerable labour could be saved. Moreover the cost for engaging labourer for taking basins/furrows etc., could be eliminated.
- Under drip system even low quality water can be filtered and used.

Management of Coconut Garden

Fertilizer recommendation for young palms: Application of fertilizers to young palm is necessary for its vigorous growth and early flowing. A general fertilizer schedule for the young palms under India conditions may be as follows:

Table 4.2: Fertilizer Recommendation for Young Palms

Months/Year after planting	*Nutrient doze*	*Fertilizer doze (gm./palms)*		
		Urea	*Super Phosphate*	*MOP*
3 months	¼ of full dose	185	250	300
2nd year	3/8 of full dose	275	375	450
3rd year	½ of full dose	370	500	600
4th year	¾ of full dose	555	750	900
5th year onwards	Full dose	740	1000	1200

Table 4.3: Fertilizer Schedule under Irrigated Condition

Months/Year after planting	*Nutrient doze*	*Fertilizer doze (gm./palms)*		
		Urea	*Super*	*MOP*
3 months	¼ of full dose	270	500	500
2nd year	½ of full dose	540	1000	1000
3rd year	¾ of full dose	810	1500	1500
4th year	Full dose*	1080	2000	2000

(Full doze = 0.5 kg. N, 0.34 kg. P_2O_5 and 1.2 kg. K_2O.)

For hybrids under irrigated conditions a fertilizer dosage of 0.68 kg. N, 0.34 kg. of P_2O_5 and 1.36 kg. of K_2O per palm/year may be required/adopted. The full doze of fertilizer could be applied in this case from 3rd year onwards as the palms tend to flower within 5 years and the initiation of flowering takes place. 32 months prior to opening of spathe. It will help in obtaining early stability in yield for the palm.

Fertilizer schedule for adult palm: In order to determine the fertilizer dozes for each location, field experiments and foliar analysis are to be adopted. If such data are not available, it is advisable to adopt the fertilizer schedule based on soil analysis and experiences with other crops: In India, manurial recommendations for all the soil groups are not available. However based on the experimental data available at CPCRI, Kasaragod, Kerala, the following general recommendations may be adopted for being tall palms.

Table: 4.4 Fertilizer Schedule for Adult Palms

	Nutrient dosage Kg./palm/year		
	Nitrogen (N)	*Phosphate(P_2O_5)*	*Potash (K_2O)*
Rainfed crop	0.34	0.17	0.68
Irrigated crop	0.50	0.34	1.20
Hybrids	0.50	0.34	1.20

In order to provide the above nutrients any one of the following fertilizers may be applied in recommended quantities for each element (NPK).

Table: 4.5 Recommended Type of Fertilizers

		Quantity in kg./year	
Nutrient	*Fertilizer*	*Rainfed tall*	*Hybrid or irrigated tall*
N	Amm. Sulphate	1.7	2.5
N	CAN	1.7	2.5
N	Urea	0.7	1.0
P_2O_5	Super Phosphate	1.0	2.0
	Ultraphos	0.5	1.0
	Mussoriphos		
	Rock phosphate		
K_2O	MOP	1.2	2.0

The above dozes of nutrients could be made available by the applications of fertilizer mixtures like 10:5:20 at the rate of 3.5 kg. /palm/year for the rain fed tall variety and 6.00 kg/yr/ palm for irrigated tall and hybrids.

When complex fertilizers are applied, it may be necessary to supplement with additional quantities of urea and MOP to have balanced dose of fertilizers as per the above recommendations.

In addition to the above fertilizers, it is also recommended to apply dolomite @ 2–3 kg./palm/year and Magnesium sulphate

@ 500 gm../palm/year especially in acid soils. All the above inorganic fertilizers are to be applied over and above the applications of organic manures like, Farm Yard Manure, compost, or green leaf manure. For gardens, which are neglected for continuous periods of 5 years or more, it is better to apply 1½ times the recommended quantity of fertilizer for the first two years of commencement of manuring.

Time of fertilizer applications: The proper time of fertilizer application depends upon the climate and soil type. The basic principle is to apply fertilizers only when the soil is sufficiently moist. In general, the fertilizer applications may be practiced just at the onset of monsoons and at the end of the rainy season. During these periods the roots will be active and leaching losses also would be minimal. In clay soils with high adsorption capacity, leaching will be less even when fertilizers are applied immediately after heavy rains.

Frequency of fertilizer application: The number of fertilizer applications should be judicious so as to ensure a continuous supply of plant nutrients to the palm. In the case of young palms, split application of fertilizers is definitely important.

In the case of adult bearing palms for obtaining better yields and more copra out turn, it is recommended to apply the total quantity of fertilizers in two split doses per year. This significantly increases the female flower production and the setting per cent. The nitrogenous and phosphate fertilizers could be applied in small doses at shorter intervals. This will also help in reducing leaching losses and soil biochemical activities thus ensuring continuous availability of nutrients for experiencing the full yield potential of the palms.

Placement of fertilizers: The placement of fertilizers should be in such a way that they are within reach of active absorptive area of the roots. The different methods adopted for the application of fertilizers are broadcasting, basin application, application in holes, circular trenches or in circular strips.

In coconuts the water and mineral absorption takes place through a small area located near the root cap. Even though a large number of roots are formed, the absorptive area is relatively restricted. In a regularly cultivated and manured coconut garden, 74 per cent of the emerging roots do not extend beyond 2 m. and 82 per cent of the roots were confined to 30–120 cm. depth of the soil. As such application of fertilizers in shallow circular trenches round the basin at a radius of 2 m. from the soil is recommended. The fertilizers can also be applied in a 90 cm.

wide circular strip 90 cm. away from the hole followed by forking and mixing the fertilizer in moist soil.

Organic manures: Application of organic manures improves the soil structure, its absorption complex and water holding capacity. When applied on an equal nutrient basis, the organic manures can alone provide the required materials to the palm and replace chemical fertilizers. Application of 25 – 50 kg./palm/year of organic manure like farm yard manure, compost, poultry dropping, green leaves, etc., supplemented with the required quantity of inorganic fertilizers is the ideal combination for obtaining vigorous growth of seedlings, early establishment and for obtaining maximum benefit.

Green manures: Majority of the soils in which coconut is grown is very poor in organic matter. Green manuring aims at incorporating organic matter with the top soil by using leguminous crops so as to augment the humus content of the soil. Nitrogen fixed by the rhizobacteria from the atmosphere in the root nosules of the leguminous cops enriches the soil with nitrogen. The green manure crops helps in uptake of mineral nutrients from subsoil and return it to the top soil by way of organic matter.

Incorporation of green manures into the soil improves the soil structure and its physical properties such as adsorptive capacity, aeration and drainage. In sandy soils with low absorption capacity, green manuring helps in conserving the soil nutrients from leaching due to heavy rains. Green manuring also minimises soil erosion and stimulates the activities of soil microflora. In gardens regularly planted with green manure crops to enrich soil with nitrogen, regular fertilizer application with expensive nitrogenous fertilizers could be reduced or done away with.

Sunhemp *(Crotalaria juncia)* wild sunhemp *(Crotalaria strata)*, Daincha, cowpea *(Vigna unguiculata)* are the most suitable green manure crops for growing in situ in coconut garden.

Both sunhemp and cowpea are quick growing succulent leguminous crops which can grow profusely even under shade of coconut plantation. Cowpea is sensitive to fungal attack. So that it is not very suitable for humid areas. Wild sunhemp withstand heavy rainfall, drought and shade. It is not browsable. Daincha thrives in shade as well an clayey and saline soils.

The green manure crops are sown with the onset of monsoons @ 30 kg./ha. In about 3 months time, they attain maximum vegetative growth and when they initiate flowering, the entire green matter produced is incorporated into the soil. About 7,000-10,000 kg. of green matter is produced per ha.

If growing of green manure crops in situ is impracticable, perennial bushes like *Gliricidea maculata* and *Tephrosia candida* can be grown on the binder, ledger and along the boundaries and the green matter produced is cut and incorporated to the soil periodically.

Gliricidia is a fast growing medium sized leguminous tree. It grows rapidly and produces large quantity of foliage within 6 months of planting, which can be periodically pruned at 30 cm. height and green matter applied to the soil during monsoons. It is propagated from seeds or cuttings. 30 kgs. leaves of gliricidia will provide sufficient nitrogen, phosphorus and potassium for coconut.

Cover cropping in coconut gardens: In coconut garden, cover cropping is the best primary control against surface run off and resulting soil erosion by breaking the impact of rain drops. Cover crops provide mulch and organic matter to the soil thereby improving the soil biological activity. Leguminous cover crops provide nitrogen. Growing of cover crops improves soil structure, infiltration of water, moisture conservation and reduces leaching of nutrients besides regulating soil temperature and smoothening weed growth.

Characteristics of suitable cover crops

a. It should be quick growing so as to cover the soil in a short time thereby controlling the weed growth.
b. The cover crop grown should dry up during drought periods so as not to compete with the main crop for moisture. It should regenerate during rainy season and be able to reproduce through seeds.
c. Cover crop should be useful as mulch through the production of large quality of green matter.

In general, cover cropping may be adopted in slopy and undulated areas where soil erosion is a major problem and in areas where effective weed control through mechanical and chemical methods are costly. In dry and drought prone areas the cover crop may compete with main crop for soil moisture and adversely affect the yield of coconuts.

The common cover crops that combine the above characters are *mimosa pudica, pueraria phaseoloides; calapogonism muconoides, centrosoma pubescens, Glyricidia maculata* can also be used as a cover crops in any soil type. At the peak of vegetative growth, the above cover crops would yield 12-24 tonnes of green matter per ha. Centrosomia however do not dry up during summer and thus competes with main crop for soil moisture. Hence this should be incorporated with the soil at the peak of vegetative growth during summer and should not be allowed to grow during summer. In young plantations, cover crop should be planted even before the seedlings are planted. Regular slashing of or dishorrowing of the cover crop is necessary to prevent it from growing too high or too dense.

Inter cultivation: Inter cultivations means working the soil in the interspaces of coconut garden by suitable implements so as to manage the soil. Different cultural techniques like shallow digging, ploughing, harrowing, etc., are found to increase the yields in coconut gardens. Ploughing helps to loosen the top soil and there by infiltration of moisture. General ploughing to a depth of 20 cm. once in a year may be adopted. In light soils, harrowing may be useful. In light sandy soils ploughing is not recommended. Piling of mounds and subsequent leveling is better in this type of soils. The effect of intercultivation mainly for the control of weeds. Intercultivation in coconut garden has been planned to conserve soil moisture status also. This may be due to cumulative respect of weed control, better percolation of water to the bottom layers, and prevention of surface run off.

Excessive inter cultivation and increasing its frequencies will result in deflection of organic matter through enhanced decomposition and destructions of the granular structure of the soil. If the soil is disturbed too deeply and frequently by ploughing, the root system may also get damaged. In general inter cultivation may be limited to weed control and for incorporating green manure with the soil.

Moisture conservation by husk burial: Even though coconut husk is commercially utilized for coir manufacture, a good per cent of husk produced are utilized for alternative uses.

Burial of husk in between vows of palms is found to be advantageous in rainfed plantations. Husks act as water reservoir in the soil (it can absorb six times its weight of water) and also supplies small amounts of potash. About 10,000 husks contain potash equivalent to one tonne of MOP. 2 kg. of husk ash would supply 1 kg. of MOP. The strong structure of husk may act as a good medium for vigorous growth of root system. The husks before burial should be stored under cover without exposure to rain water as otherwise half of the potash content in them will be lost by leaching. The beneficial aspects of husk burial in terms of yield increase will be noticed from third year and effect would last for a period of 6 years, after which the process is to be repeated. Husks should not be buried in coconut garden subject to water stagnations as it will worsen the situation.

Generally husks can be buried during the beginning of the monsoons @ 1000 husks/tree in linear trenches in between rows of palms. The trenches may be 2 m. wide and 40 cm. deep and they have to be taken against the slope. The husks are buried with the spongy linear surface facing upwards. The husks are covered with soil and the process can be continued until the layer is just about 20 cm. below ground level.

The husks can also be used as surface mulch around the base of the palm. In this case, husks are placed with the convex side up round the palm upto a radius of 2 m. from the base. Coir pith or coir dust can also be incorporated with the soil as mulch, which will help improving soil structure and its physical properties.

Organic manure from coir pith: Coir pith contains very less nitrogen and has large amounts of lignin and phytotoxic polyphenols and has to be composted before using as a manure. Exposure to rains and sunlight for many years results in loss of problematic chemicals and the use of weathered coir pith may be advantages. Fresh coir pith has a wide C:N ratio (about 100:1) and for initiating microbial action, nitrogenous organic or inorganic materials are to be added. Additionally, fortification with rock phosphate at the rate of 10 kg. per tonne of coir pith can also favour microbial action. The well known technology for composting of coir pith using *Pleurotus sajor caju* may be utilized for large scale composting. For composting one tonne of coir pith, 5 kg. urea and 5 bottles of Pleurotus spawn are required. Hundred kg. of coir pith is spread on a leveled land in a shaded place and one bottle of spawn is sprinkled over it. The spawn layer is covered with 100 kg. coir pith and 1 kg. urea is sprinkled over it. This process is repeated five times to get a heap, which is protected from direct sun light and rain. Proper moisture is to be maintained in the heap and is allowed to undergo degradation for one month. This composted coir pith can be used as manure in coconut plantations.

Weed control in coconut gardens: Necessity for weeding depends upon the soils and climate, the weed species and age of the plantation. In young plantation, weeding is essential. Ring weeding in a circle with a diameter of 75 cm. and widening with the age of the palm is usually practiced in coconut gardens. Weeding may be carried out once a year or more depending on local condition.

There are 3 different methods of weeding; manual, mechanical and chemical. Good results can be obtained with all the three methods but the local condition will determine the relative merit and advantage of one method over the other.

Intensive weeding should be restricted to a circle around the palm. Manual hoeing and deep mechanical harrowing cause depletion of the soil. Chemical weed control using herbicides like Paraquet, Divan Diuran are used elsewhere to control weeds in young plantations. These are sprayed in a circle with a radius of 1.65 m. to 2.4 m. around the palms. When chemical herbicides are used, care should be taken not to hit the coconut leaves while spraying. However the effectivity and economy of different methods of weeding is still a controversial matter.

In general weeding is most effective when palms grow under moisture and nutrient deficiency. Weeding around young seedlings is always necessary. When the cost of weeding is prohibitive, it is always better to adopt intercropping with economically viable intercrops.

Integrated Pest management in Coconut

The major insect pests of the coconut palm are the rhinoceros beetle *(Oryctes rhinoceros)*, the leaf eating caterpillar *(Opisina arenosella)*, the red palm weevil *(Rhynchophorus ferrugineus)* and the root eating white grub *(Leucopholis coneophora)* (Nampoothiri 1999).

Rhinoceros beetle: This is the most serious pest, which has an ubiquitous distribution. The adult beetle bores through into the unopened fronds and spathes. The affected frond when fully opened shows the characteristic geometric cuts. Infestation on spathes often destroys the inflorescence and thus prevents production of nut. The beetle breeds in a variety of materials such as decaying organic debris, farm yard manure, dead coconut stumps, logs and compost. The total duration of life cycle of this pest is about six months.

Maintenance of sanitation in coconut gardens by proper disposal of decaying organic debris is an important step in the management or rhinoceros beetle. Mechanical method of control is possible by extracting beetles with hooks, without causing any further injury to the growing point of the palm.

The innermost 2nd and 3rd leaf axil may be filled with Sevidol G (25 gm.) + fine sand (200 gm.) per palm. Filling lead axils with 10.5 gm. Napthalene balls covered with sand at 45 days interval is also effective. The breeding sites of beetle such as cattle dung, compost and other decaying organic debris may be treated with Carbaryl 50 per cent dust or WP at 0.01 per cent. Three applications in April, September and December are adequate to give sufficient protection to palms in heavily infested tracts.

Baculovirus inoculated beetles @ 10-15 per ha. can be released to bring down the pest population. Release of the exotic predator Platymeris leavicollis was also found to be effective. Inoculations of breeding ground with entomopathogen, *Metarhizium anisopliae* (Metch) @ 5X10" spores gave effective control. Adoption of all the proven methods of beetle control in an integrated manner would give quicker reduction of infestation.

Leaf eating caterpillar: Leaf eating caterpillar is another serious pest of coconut in the coastal and backwater tracts. In recent years, this pest got access into certain interior tracts as well as assumed severe proportions. The caterpillars live under surface of leaflets, inside silken galleries and feed voraciously on the chlorophyll containing functional tissues. This affects the health of the palm adversely and results in reduction of yield. The severity of infestation by this pest will be marked during the summer months from February to June. With the onset of south west monsoon, the pest population begins to decline.

Spraying of infested palms on the lower surface of leaves with Dichlorvos (0.02 per cent) would give satisfactory control of the pest. Chemical control treatment is generally adopted only in cases of severe outbreak of the pest. The treatment may be done at quarterly intervals in March, June, September and December depending on the abundance of the pest in the field.

As this pest is subject to parasitisation by a good number of indigenous larval and pupal prasites, its biological suppression is feasible. Mass multiplication, liberation and colonization of indigenous and/or exotic parasites like *Parasierola nephantidis, Elasmus nephantidis, Brachymeria nosatoi* and *Xanthopimpla punctata* in the infested fields would control the pest population. The parasite breeding stations established in the major pest-infested regions are mass culturing and releasing different species of larval and pupal parasites of the pest. The bio-control laboratory at CPCRI maintains culture of important species of parasites for supply as nucleus culture material to the parasites breeding stations in the country for mass multiplication and release in infested gardens.

Red palm weevil: Red palm weevil is the most dreaded pest of young coconut palms. Generally palms of the age group 5-20 years are affected by this pest. At times, seedlings below 5 years are attacked by the pest. Since the pest is a tissue borer, its detection in early stage of infestation is rather difficult. The major diagnostic symptoms of red palm weevil infestation are the presence of holes in stems, oozing viscous brown fluid and extrusion of chewed up fibres through the holes, longitudinal splitting of lead bases and wilting of inner leaves. Sometimes the gnawing sound produced by the grubs feeding inside will also be audible. Quite often the infestation would become evident only when the growing point of the palm is damaged and the crown has topped.

Affected palms can be saved by injection of Carbaryl (Sevin) at 1 per cent concentration. 20 gm. of 50 per cent Sevin in 1 litre water per palm should be introduced into the trunk through a hole above the infested portion, using an auger and funnel. All the holes on the affected stem should be plugged after injecting the insecticide suspension to the palm. If the pest infestation is through the crown, the insecticide suspension should be slowly poured in after cleaning the crown of all affected materials. Endosulfan 0.1 per cent is also found to be effective in controlling this pest.

Coconut logs of 50 cm. long, split longitudinally and cut surfaces smeared with fresh toddy fermented with yeast and acetic acid are effective traps. Weevil thus trapped can be collected and killed. Pealed coconut petioles arranged in trays and treated with mascerated sugarcane + yeast are suitable substitutes for coconut log traps.

Entry of this pest through the cut ends of lead base can be prevented by leaving a length of 120 cm. of petiole while cutting leaves, particularly from palms of the susceptible age. Cutting steps on the stem for easy climbing also has to be discouraged as a precautionary measure to avoid entry of the pest through injured tissues. Palms affected by leaf rot / bud rot diseases are more prone to weevil infestation. As such, they are to be treated with an insecticide as well, after the fungicidal treatment. Dead palms should be cut and burnt. An integrated approach involving all the proven methods of weevil control is feasible in maintaining gardens free of pest infestation.

Root eating cockchafer: The soil inhabiting white grubs cause damage to the roots of coconut. Besides coconut, it infests tuber crops like tapioca, colocasia, sweet potato etc., grown as intercrops in coconut gardens. The leaves of affected palms become sickly pale yellow. In cases of heavy infestation, there will be immature nutfall as well.

Application of Phorate 10 G @ 100 gm. per palm in May – June and September – October so as to cover pre and post monsoon period is recommended in coastal and Malanad areas. The pesticide applied should be mixed and raked in the top 15 cm. soil, so as to give effective control of the pest.

Eriophyid mite: Coconut cultivation in India is now threatened by the nut infesting eriophyid mite Aceria guerreronis (Keifer). This notorious pest on coconut was first reported from India in 1998 from Ernakulam district of Kerala State. The mite has assumed severe proportion subsequently and is now known to cause considerable crop loss in Kerala, Tamil Nadu, Karnataka, Pondicherry parts of Andhra Pradesh, Lakshadweep, Goa and Orrisa (CPCRI, 2002).

The first report of A. gurreronis on coconut was in 1960 from Mexico. Later its occurrence was widely noticed in several countries in South America and Carribean region. In 1981, severe damage by pest was reported from the western parts of Africa and in 1999 from Sri Lanka.

The mites take about 7-10 days to complete their life cycle. Under favourable conditions, they multiply and spread rapidly. The dispersal of the pest is mainly through wind. Honey bees and other insects visiting inflorescence of coconut also act as agents for dispersal of mites. The female mite has a high fecundity and lays about 200 eggs on an average. Various states of the mite are seen in the floral bracts and the tender portion of developing nuts covered by the perianth. Colonies of the mite harbour on immature nuts of one to five month growth. The unfertilized female flowers do not harbour any mite.

The eriophyid mites are microscopic having an elongated worm-like body. The adults reach a size of 250 micron (about one fourth of a millimeter) on an average. Their bodies are finely ringed and bear two pairs of legs on the anterior end.

Symptoms of attack

The mites suck the plant sap with their needle like mouth parts, living on the tender meristematic region covered by the perianth. The first symptom of mite attack is the appearance of elongated white lines below the perianth. Within a short time these feeding marks appear as pale yellow triangular patches just below the perianth and gradually turns brownish in colour. As the nuts grow, the injury forms warting and longitudinal fissures on the nut surface. Severe infestation by the mite results in poor development of nuts with reduced kernel weight and poor quality of fibre.

Severe infestation by the mite during early stages of nut development results in excessive button shedding. Heavy yield loss is observed, along with reduction in fibre content.

Management of the pest: With the present know-how, it becomes necessary to adopt chemical control measures with pesticidal spraying to manage the mite. Studies are in progress on the possibility of bio control of the mite using predacious mites/insects and pathogenic fungus like Hirsutella thompsonii. Investigations on the management of the pest with various pesticides carried out at CPCRI have proved the effectiveness of the following pesticides.

Botanical pesticides: Spraying of affected palms with a mixture of 2 percent neem oil, garlic and soap mixture has been found to be effective. To prepare one litre of the above mixture, 20 ml. neem oil, 20 gm. of cleaned garlic pearls and 5 gm. washing soap are required. Dissolve the soap in 500 ml. of water and add the neem oil to this solution and mix it well so that it forms an emulsion. In another, 500 ml. water mix the well ground garlic and add this to the soap neem oil mixture by sieving through a cloth to remove the debris of garlic pearls. This mixture is stirred well and thus can be used for spraying. The same shall be used on the day of preparation.

Azadirachtin, the active insecticidal principle in neem oil is commercially made and used as an insecticide. Studies have proved that 0.004 per cent Azadirachtin gives very good control of the pest in the field. Spraying of botanical pesticide containing 1 per cent Azadirachtin at the rate of 4 ml. per litre of water is recommended for the management of the mite.

Chemical pesticides: Various chemical pesticides have also been found to be effective for the mite management. Sulphur is recommended for spraying at 0.4 per cent concentration for mite control (5 gm. per litre of water). Spraying of either 0.2 percent Triazophos/0.1 per cent Dicofol / 0.05 per cent Carbosulfan has also been recommended for the effective management of the mite based on field trials.

Integrated Disease Management in Coconut

Disease that occurs on coconut palms account for substantial loss in their productivity. Among the maladies, bud rot, root (wilt), leaf rot, Thanjavur wilt, stem bleeding and Tatipaka are important as they cause significant decline in yield and in certain cases the death of the palm. (Nampoothiri et. al., 2000)

Bud rot: It is caused by the fungus: *Phytophthora palmivora*. The earliest symptom is the yellowing of one or two younger leaves surrounding the spindle. The spindle withers and droops down. The tender leaf base and soft tissues of the crown rot into a slimy mass of decayed material emitting a foul smell. The disease kills the palm if not controlled at the early stages. Palms of all age are liable to be affected but normally younger palms are more susceptible. The disease is more prevalent during the monsoon months when the temperature is low and humidity is high.

Control: In early stages of the disease, when the spindle leaf starts withering, all affected tissues of the crown are to be cut and removed and Bordeaux paste is to be applied which has to be protected from rain by providing a polythene covering till normal shoot emerges. All disease affected tissues should be removed from the palm and burnt. As a prophylactic measure, 1 per cent Bordeaux mixture is to be sprayed on spindle leaves and crown of affected palms as well as neighbouring palms before the onset of monsoon. Leaf axil filling with Sevidol 8G, 25 gm. mixed with 200 gm. sand is also recommended to prevent red palm weevil infestation of affected palms.

Dwarf varieties of coconut are sensitive to copper injury. Hence Bordeaux mixture spray should not be done; instead, small perforated sachets containing 2-3g. of Indofil M-45 may be placed in the top two or three leaf axils to control bud rot disease.

Root (wilt) disease: The root (wilt) disease of coconut made its first appearance in Kerala after the great floods in 1882. Since then the disease has spread to many places. Presently the disease occurs in a continuous manner in eight southern districts of Kerala. Root (wilt) disease is also observed in a sparse manner in isolated pockets in the remaining six northern district of Kerala and also in a few locations in the neighbouring state of Tamilnadu.

The causal organism is Phytoplasma. The disease is transmitted by the lace bug *Stephanitis typicus* and plant hopper *Proutista moesta*.

Symptoms: Flaccidity, the characteristic bending or ribbing of the leaves, general yellowing and marginal necrosis of the leaflets are the important visual diagnostic symptoms. Nuts are smaller and kernel becomes thin. Oil content of copra is also reduced.

Management: Root (wilt) disease is not lethal, but debilitating nature. No curative measure has so far been identified against it. The strategy is to contain the disease to its present geographical limits and managing the disease by improving the condition of affected palms and increasing the yield through proper manuring and other agronomic practices such as, eradication of the disease in mildly affected areas by cutting and removal of affected palms; in the heavily disease affected tracts, removing all the severely affected uneconomic adult palms (those yielding less than 10 nuts per palm per year) and all diseased palms in the prebearing age;

Application of the recommended dose of NPK, 3 kg. Magnesium sulphate and 50 kg. organic manure per palm per year; organic recycling by following mixed farming system; growing suitable inter/mixed crops; basin management with green manure crops; irrigation during summer months and control of leaf rot disease which is usually noticed in root (wilt) affected palms. Replanting with progenies of disease free palms located in hot spot areas enhances the yield of the palms.

Leaf rot: Leaf rot disease commonly occurs on coconut palms already affected by root (wilt) disease. Infection by this disease is the major reason for the low productivity of root (wilt) affected palms. *Collerortichum gloeosporioides, Exserohilum rostratum* and Fusarium solani are the fungal organisms causing leaf rot.

Symptoms: The first visible symptom of the disease is blackening and shriveling up of the distal ends of the leaflets in the central spindle and in some of the younger leaves. Later the affected portion breaks off in bits giving the infected leaves a fan like appearance. If no protective measures are undertaken, each new leaf of the diseased tree gets infected with the result that a stage is soon reached when all leaves of the tree show disease symptoms.

Control: The rotten portion of the spindle and two successive leaves are to be cut and removed; 300 ml of fungicidal solution containing 2 ml of Hexaconasol (Contaf 5EC) or 3 gm. Mancozeb (Indofil M-45) may be poured in the cavity around the base of the spindle; 20 gm. Phorate 10 G mixed with 200 gm. fine sand to be applied around the base of the spindle to ward off insect pests and the palms are to be treated twice in a year i.e., during April-May and October-November. Palms in the early stages of disease will recover totally with two or three applications. Palms in the advanced stages (with an index of more than 50 per cent) would take three years to recover fully. To prevent the reoccurrence of the disease, the treatment needs to be continued.

Stem bleeding disease: Stem bleeding disease is prevalent in all the major coconut growing states in India. The fungus, Thielaviopsis paradoxa is the primary causative agent. Growth cracks on the trunk, severe summer followed by sudden wetting, imbalanced nutrition, excess salinity etc., are the predisposing factors.

Symptoms The disease is characterized by the exudation of dark reddish brown liquid from the longitudinal cracks in the back, generally at the base of the trunk. The bleeding patches spread throughout as the disease advances. The liquid oozing out dries up and turns black. The tissues below the lesions become rotten and turn yellow first and later black. Leaves in the outer whorl turns yellow rather prematurely droop and dry. Production of bunches is affected. Nut fall also is noticed. The trunk gradually tapers at the apex and crown size becomes reduced.

Control: The affected tissues are to be completly removed using a chisel and dress the wound with Calixin 5 per cent and apply coal tar after 1-2 days. Root feeding of 100 ml. Calixin 5 per cent has to be done thrice a year during June, October and January prevents further spread of lesions. Recommended dose of fertilizers are to be applied and irrigation provided during summer. 5 kg. neem cake per palm has to be applied during September–October.

Thanjavur wilt / Ganoderma disease: Thanjavur wilt first appeared in the coastal areas of Thanjavur district of Tamil Nadu after cyclones of 1952 and 1955 and hence the name Thanjavur wilt. Now it is present in all the coconut growing areas in Tamil Nadu and in some part of Kerala. The Fungi, Ganoderma lucidum and Ganoderama applanatum are the causative agents of the disease.

Symptoms: Decay of root system, flaccidity of spindle leaves, bronzing of outer leaves, arrested fruit set and appearance of bleeding patches on the basal region on the stem are the symptoms observed. Ultimately the palm dies off, in advanced stages, the bracket of fungus causing the disease is seen on the stump.

Management: Organic manures (50 kg.) and 5 kg. neem cake is applied per palm and irrigation is to be provided. Drainage channels are to be provided between rows of palms. The affected palm are to be isolated from the healthy ones by digging a trench around the affected palm; Phytosanitary measures viz., removal of dead palms, burial of the affected roots and bole in a pit, etc., are to be carried out; 100ml. Calixin 5 per cent may be applied through root at quarterly intervals for one year and the soil has to be drenched with 0.1 per cent Calixin @ 25 lit. per palm.

Mahali (Fruit rot and nut fall): The fungus, Phytophthora palmivora causes the disease. Shedding of female flowers and immature nuts are the symptoms of the disease. Lesions appear on the young fruits or buttons near the stalk which later develop into decay of the underlying tissues.

Spraying 1 per cent Bordeaux mixture or 0.5 per cent Copper oxychloride preparations on the crown of palms once before the monsoon and once or twice after the monsoon at 40 days interval to control the disease. The fallen (disease affected) nuts are to be collected and burnt.

Leaf blight or Grey leaf spot: Fungus, *Pestalotia palamarum* causes the disease. In the mature leaves of the outer whorl, yellow specks encircled by a graying band appear which later turn to grayish white. The sports coalise into irregular necrotic patches causing extensive leaf blight. When the infection is severe, the leaf blade completely dries and shrives off. Cutting and removal of older affected leaves and spraying the foliage with 1 per cent Bordeaux mixture controls the disease.

Tatipaka disease; This is named after its place of occurrence in, the Tatipaka village of East Godavari district in Andhra Pradesh. It is a slow spreading disease and affects mostly heavy bearing palms of middle age group. At present, the incidence level of this disease is reported to be negligible. Electron microscopic evidence supports a phytoplasmal etiology for the disease.

Symptoms: Development of an abnormally large crown with dark green inner leaves and higher yield is the precursor of disease incidence. subsequently, the crown becomes smaller in size producing progressively shorter leaves. The stem begins to taper. The leaves give a fascinated appearance due to improper unfolding of leaflets. The affected tree produces smaller bunches with atrophied barren nuts.

Infected trees cannot be cured. Spread of the disease can be prevented by strictly prohibiting the use of planting materials from the affected areas.

Crown choke disease: Crown choke symptom is caused by acute shortage of boron. The disease is commonly observed in Assam and West Bengal.

Symptoms: Emergence of shorter leaves with deformed and

crinkled leaflets is the first symptom. Often the leaflets show severe tip necrosis and fail to unfurl. In many cases, it shows a choked appearance to the frond. As the disease progresses, stick like leaf stalks emerge.

Soil application of borax @ 50 gm. per palm at half yearly interval i.e., during February-March and August-September improves the condition of the palm. Making slits in the tight extension of the petiole and opening out the crown would help in the normal emergence of fronds.

Coconut based Farming Systems

Inter cropping: The practice of growing annual or other short duration crop in the interspaces of a perennial crop is called inter cropping. (Eg. Tubers, rhizomes, vegetables etc., intercropped in coconut garden).

Mixed cropping: The practices of growing a long duration crop or a perennial crop in the inter places of another perennial crop is called mixed cropping. Eg. Cacao, nut meg, clove etc., in inter places of coconut gardens.

Multi-storied cropping / multiple cropping: The practice of growing, different species of compatible crops having varied morphological and physiological features, in the interspaces of a perennial crop (coconut) is called multi-species cropping. In this intensive mixed cropping system, the crop canopies intercepts solar radiation at varying heights and the roots also anchorage at varying depths. Eg. Growing of pepper, cacao, pineapple, etc., in coconuts gardens.

Mixed farming: Mixed farming is a multiple cropping system wherein livestock or poultry are profitably integrated with coconut by growing fodder/pastures and legumes in the interspaces of palm. Fodder grasses can be used for maintaining milch animals. Cow dung can be used for producing biogas to meet the fuel and lighting requirements of the farmer and the digested biogas slurry can be recycled as manure for the coconut – fodder cropping system.

High density multi-species cropping system: This is a multiple cropping system developed by CPCRI wherein several plant species both annuals and perennials are grown together in a definite pattern to accommodate maximum number of plants per unit area of land, so as to utilize the natural resources of soil, air space and sunlight for the production of bio mass having high economic value and diverse end use. (Example: Black pepper, clove, banana yam, turmeric etc., in coconut *plantations*).

Prospects of Coconut based Farming Systems in India

Coconut being a small holders' crop under mono-cropping it neither provides gainful employment opportunities for the family labour (150 man days/ ha./year) nor generates sufficient income. Moreover, in recent years many coconut growers have suffered economically due to unstable prices (CDB, 2001).

In coconut holdings with palms spaced at 7.5 X 7.5 m., nearly 75 per cent of the land is unutilized by the palm. Maximum root activity is in the area around the base of the palm up to a radius of 2 m. and to a depth of 1.2 m. The growth habit and canopy configuration is such that up to 8 years after planting and also after 20-25 years, the light infiltration to the ground floor is adequate (about 70 per cent) to raise a variety of crops. During the 8-25 years period, the canopy is developing but the palms are tall enough to allow sufficient incident light and this creates less than ideal condition for cropping the interspaces.

Under such a situation multi-species cropping systems ensure maximum resources capture and use, leading to higher yields per unit area of soil, water and light. Nevertheless only a few farmers practices intercropping/mixed cropping scientifically in coconut gardens. Cropping systems involving different compatible annual or perennial crops in the interspaces, ensure additional employment and income. The improvement in the soil properties in general and biological activities in the root region in particular, result in the modification of soil environment to the benefit of plant growth.

Inter Crops

Various tuber crops, rhizomatous crops, cereals, vegetables, pulses, oil seeds, fruit crops, flowers and medicinal and aromatic plants have been found to be suitable for raising as intercrops in coconut garden (Nambiar, 1978, Verghese et. al., 1978, Gopalasundaram and Nelliat, 1979, Rethinam and Nampoothiri et. al., 2000).

Tuber crops: Tapioca, elephant foot yam, sweet potato, colocasia, greater yam and lesser yam are the tropical tuber crops cultivated as intercrops in coconut garden. The tuber crops partially meet the food requirements of a family and are always found in the homestead gardens.

Rhizome/spice crops: Ginger and turmeric are the important rhizome spice crops commonly intercropped in coconut gardens. They are planted in raised beds of 1-1.2m. width, 30 cm. height and convenient length. Farm yard manure is to be mixed with soil or the seed material is to be covered with farm yard manure in pits after planting. Ginger rhizome bits are planted at a spacing of 25 cm. X 25 cm. at a depth of 4.5m. with bud facing upwards. Turmeric finger rhizomes are planted at spacing of 15 cm. X 30 cm. covered with dry powered cattle manure. Mulching with green leaves is to be done immediately after planting which is to be repeated after 50 days and 100 days of planting. Ginger is highly susceptible to soft rot disease against which proper plant protection measures are to be taken to ensure a good crop. The yield of ginger and turmeric as intercrop under average management conditions ranges from 7,000 to 8,000 kg./ha.

Cereals: Studies at Kasargod revealed that upland rice varieties perform fairly well as intercrop. The rice variety 'Rohini' is the best performing variety. The upland rice can be sown immediately after the onset of monsoon.

Maize *(Zea mays)*, varagu *(Paspalam scrobiculatum)*, pearl millet *(Pennisetum typhoides)* and finger millet *(Eleusine coracana)* perform well as intercrops in coconut garden.

Vegetables: Raising of chillies, potato and French beans is found to be a profitable practice in maidan areas of Karnataka. The experiments conducted at Kasaragod have indicated that vegetables like snake gourd, bottle gourd, amaranthus, coccinia, brinjal and bitter gourd are compatible intercrops with coconut. Among different crop sequences tried, snake gourd – ridge gourd– amaranthus was found to be the most remunerative followed by amaranthus – bottle gourd – brinjal.

Pulses: Leguminous pulse crops help in enriching the soil with nitrogen when grown in association with coconut. Among the pulse crops, cowpea, blackgram, green gram, bengal gram, short duration red gram and soya bean give satisfactory yield in the maidan areas of Karnataka. Soya been varieties PK-472, MACS-13 and MESCS-48 are highly promising as intercrop in coconut garden. Under Kerala conditions, horse gram, cow pea, green gram and black gram are the suitable intercrops giving yields to the tune of 250-500 kg. per ha.

Oil seeds: Oil seed crops like ground nut are remunerative intercrops in coconut garden which also enrich the soil.

Fruit crops: Banana and pineapple are the most popular fruit crops grown as intercrops in coconut garden. In Kerala, Palayankodan, Robusta, Karpooravally and Poovan are identified as the suitable banana varieties for intercropping which give an yield of 8-12.8 kg./bunch. Pineapple as an intercrop performs well in the partially shaded conditions of coconut gardens. A mean yield of 1.50 kg. per fruit was obtained.

Floriculture: Orchids, anthuriums and other cut flowers and ornamentals can be successfully grown as intercrop in coconut garden.

Medicinal and aromatic plants: Presently there is high demand for natural herbs used in the preparation of ayurvedic medicines. Crops such as lemon grass, kacholam, dioscorea, arrowroot, sida, thippili (long pepper), neela amari and adapathiyan are suitable as intercrops in coconut.

Mixed Crops

Various perennial crops have been tried in the coconut gardens especially in Kerala, Tamilnadu and recently in Andra Pradesh. The suitability of crops like cocoa, pepper, clove, nutmeg and cinnamon as mixed crop in coconut plantations have been reported by several authors (Bavappa, 1975, Nair and Gopasundaram et. al., 1990, Rethinam and Venugopal, 1994 and Das, 1999).

Cocoa: Cocoa is an ideal mixed crop for coconut garden. One year old cocoa grafts are planted at a spacing of 3m. between plants in a single row system in between two rows of coconut palms. In a coconut garden with palms spaced at 7.5m. X 7.5m., about 450 cocoa plants can be accommodated. Double hedge system can be followed for cocoa planting in coconut gardens with wider spacing i.e., 8.5 m. and above. It gives stabilized yield from seventh year onwards. On an average 10-15 kg. of fresh pods per tree are obtained every year. 10-12 kg. pods give 1.0 kg. wet beans and 3.0 kg. of wet beans give 1.0 kg. of dry beans.

Pepper: Black pepper is commonly raised as a remunerative mixed crop with coconut in the west coast of India. Pepper varieties viz., Karimunda, Panniyur-2 and Panniyur-5 are found to be suitable for mixed cropping. It can be planted in the coconut basin using coconut trunk itself as the standard and also in the inter space. Rooted cuttings of pepper may be planted at a distance of 1m. from the bole of the palm on the northern side in pits of 50cm. x 50cm. x 50cm. size. The vines may be trained along the ground and then on the palms by tying to the trunk during the first two years. The height of the vines is to be restricted to 4m. by pruning. Pepper begins to yield from third year onwards and comes to stabilized yield by 7-8 years. On an average, one kg. of dry pepper can be obtained from a vine.

Clove: Clove can be grown as a remunerative mixed crop in coconut gardens in fertile, well drained soils with assured irrigation. It is planted at the centre of four coconut palms. Clove starts flowering at the age of six years and full bearing comes at 20 years. The flower buds are harvested when they turn light pink from green by hand picking with the help of plant form ladders. Flower buds are dried in sun for 4-5 days till they turn brown in colour. An average yield of 3 kg. is obtained from a tree per year.

Nutmeg: Two year old nutmeg grafts are planted in the centre of four coconut palms. It is preferable to use grafts prepared from high yielding female trees because of the dioeceous nature of the plant. While planting, the ratio between male and female grafts is to be kept a 10:1. Nutmeg flowers at 5-8 years of age and full bearing comes at 15-20 years. Fruits are to be harvested when they have split and the aril turn bright red in colour. The mace is dried in the sun for 10-15 days, till they become brittle and turn yellowish brown from the initial red colour. The nuts are dried till the kernel rattles within the shell. On an average, 1500-2000 fruits per tree will be obtained which comes to 8-12 kg. nuts and 1.5-2.0 kg. mace.

Cinnamon: Cinnamon is another tree spice, which comes up well as a mixed crop. It can be planted in the double hedge system at a spacing of 3 m. between plants. Harvesting can be started from the fourth year and continued in alternate years. Shoots of finger thickness and uniform brown colour are cut and the bark is extracted by peeling. Harvesting is a labour intensive operation and hence mixed cropping with cinnamon will be profitable only when family labour is used or labour is cheaply available.

Multistoried Cropping

A four crop combination of coconut, pepper, cocoa/cinnamon (single and double hedge) and pineapple was tried in a 20-year old coconut plantation (Nelliat, 1979). The spread of the above ground parts of cocoa and cinnamon which are pruned periodically and confined to heights less than 3.5 m. This constituted first floor. Pepper vines having its canopy from 2 to 8 m. on the coconut trunk forms the second floor. Pineapple forms the ground floor in the system. In this, feasibility trial initiated in 1979, the mean yield of coconut increased by 33 per cent. The output from the four crops multistoried cropping were 17,000 coconuts, 300 kg. cocoa (dried beans), 60 kg. dry pepper and 4,000 kg. of pineapple per ha. per year.

Mixed Farming

Fodder grasses like hybrid napier, guinea grass and Guatemala grass yield about 50 to 60 tonnes of green fodder per ha. per year under coconut shade which is sufficient to maintain five crossbred milch cows. Leguminous crops like Brazilian lucerne, cowpea and pueraria also perform well. Summer irrigation is to be given using sprinklers to ensure a steady growth of fodder through the year. Grasses are to be replanted every four years after giving a through digging to prevent root matting. A mixed farming model generated additional employment to the tune of 850 mandays and ensured good returns without any yield decline in coconut (Mathew and Shafee, 1979). Nair and Sahasranaman (1978) reported good forage production potential in irrigated coconut gardens. Mixed farming system in 1.04 ha. coconut garden yield 11,276 coconuts, 60 kg. pepper, 250 kg. banana and 14,495 litres of milk. The total revenue from this mixed farming system was Rs.67,705 and the net income from the cropping system was Rs.43,654.

Most of the annuals, biennials and perennials tried in coconut-based multiple cropping systems are compatible with coconuts. Not only were most of them found beneficial to coconut productivity because of consequential site enrichment, but also their productivity was often comparable to that when grown in the open. Hence, the coconut-based cropping systems are capable of improving the financial status of smallholders, while permitting them to use available resources more efficiently. But the success of these systems depends on the choice of component species and the availability of suitable shade-tolerant varieties of these species, time of planting, labour resources, access to capital and market outlets, price behaviour, availability of irrigation facilities in dry season and the theft problem. Nevertheless, the crop mix is better for smallholder coconut farming than monocropping (Das 1999).

Harvesting and Post Harvest Processing

Harvest

- Coconuts are harvested from a tree at varying intervals in a year.
- The frequency differs in different localities.
- In the West Coast, nuts are generally harvested from six to twelve times a year.

- In a good well-maintained garden, bunches are regularly produced, and harvesting is done once every month when the nuts are mature.
- In the poor soils of the laterite type, there may be only six harvests.
- In the coconut tracts of Bombay, generally four to six pickings are taken in the year while in Bengal, Orissa and Assam, harvesting is usually done twice a year.
- It is a good practice to harvest coconuts only when they are fully mature, unless for special markets less mature nuts are required.
- Tender coconuts are picked as and when required.
- Coconuts become mature in about twelve months after the opening of the inflorescence or flower bunches.
- In a regularly and heavy bearing tree, bunches appear and mature normally at intervals of about thirty days and a picking can be taken every month.
- When the interval between two harvests is long, two or three bunches (twelve, eleven, and ten months old) are sometimes harvested from a tree at a time.
- This is especially common in the West Coast where green husks are in great demand for the manufacture of coir.
- The husks of fully mature nuts become dry in a short period after harvest and dry husks are not generally used for the manufacture of coir as they give inferior fibre as compared to the fibre obtained from less mature or green nuts, i.e., nuts which are eleven and ten months old.
- Harvest of nuts which are less than eleven months old has been found to be definitely uneconomic for the production of copra and oil, as there is considerable loss in the quality and quantity of these products in the less mature nuts.
- Nuts which are only eleven months old give fibre of good quality and can be harvested in the tracts were green husks are required for the manufacture of coir fibre.
- The slight reduction in the quantity of copra and oil will be easily compensated by the additional income from the fibre.

Method of Harvest

- Reaping the nuts which have been shed by the palm is

the simplest way of harvesting coconuts but the yield of nuts produced is likely to be reduced.

- Harvesting of coconuts is commonly done by climbing the tree with the help of a rope ring round the feet or ankles of the climber or by using a ladder.
- On reaching the top, the climber taps the nut in the lowermost bunch with its harvesting knife to test its maturity.
- If he is satisfied, he cuts the bunch at the base of the stalk when it drops down to the ground.
- A climber may climb about 40 – 50 palm per day and depending upon the productivity of the palm and length of the period between climbing rounds each climber may harvest 400 – 800 nuts per day.
- If the ground is very hard or if tender nuts are to be harvested, the bunches are lowered by using a rope. The climber also cleans the crown and removes the dry leaves, sheaths and spathes.
- In the West Coast and certain other tracts where coconut leaves are required for thatching houses, one or two lowermost leaves are also cut down at the time of harvest.
- The cutting down of green leaves is considered undesirable as it affects the yield of trees to some extent.
- In some places where the trees are not tall, harvesting is done by cutting the bunches with a knife, attached to a long bamboo pole.
- Nuts which are to be stored for making ball copra are not harvested till they are completely ripe and dry.
- In Malaysia, Indonesia and Sri Lanka harvesting is done with bamboo poles with a knife attached to the end to cut the bunches. This method can be employed in homesteads and in dwarf palms. By this method, 250 trees could be harvested per day.
- In Malaysia, Thailand and Indonesia, the big tailed macaque *(macacus nemestrina)* is trained to harvest of the coconuts. Efficient monkeys may harvest upto 1000 nuts in one morning.

Storage

- In India, the harvested nuts, particularly those that are to be converted into copra are stored for some time in shade in specially constructed coconut stores.

- If, however, the husks are required for retting, the nuts are husked soon after harvest.
- In Ceylon, the nuts are kept heaped in the open garden till they are disposed off.
- A few nuts in these heaps are often found to germinate or rot.
- From the point of time of view of making copra a limited period of storage of unhusked nuts is found to be beneficial in the following ways
- Husking is made easier as fresh green nuts are more difficult to husk.
- Shelling is cleaner and easier since the kernel detaches quicker from the shell.
- The resulting shells are dry hard and burn well without smoke.

Yield

- Under favourable conditions, coconut palms commence to flower in about five to six years after planting.
- In loamy soils, it may be in seven to ten years and in poor soils it may be as late as fifteen to twenty years after planting.
- In about five years after the commencement of the flowering, the trees begin to yield normally.
- Proper manuring, intercultivation and irrigation are known to induce early flowering and bearing.
- The yield of nuts from a tree varies considerably, depending upon a number of factors such as the soil, rainfall and seasonal conditions, the variety planted, the age of the trees, manuring, intercultivation and the effect of pests and diseases.

Properly maintained and regularly cultivated and manured gardens in loamy soils under unirrigated conditions yield about sixty to hundred nuts per tree per year, while it would be definitely more in the rich alluvial soils. In poor soils and in neglected gardens, the yield may be twenty nuts or less per tree per year.

Seasoning of Harvested Nuts

The harvested nuts are generally heaped under shade for a few days so that the moisture content of kernel decreases and the copra thickness increases, even when nuts are harvested during

9th – 10th month. Storage helps in increasing the copra out turn but spoilage on storage may be more. In fully matured nuts the quality of copra will be superior if it stored for few days under shade immediately after harvest. Storage also helps easy husking and for obtaining hard, dry shells: Studies conducted in India showed that when stored for 60 days the copra content of 9 months old nuts increased from 96.8 to 127 gm. in 10 months old nuts from 128.2 to 138.9 gm., and in 12 months old nuts from 161.1 to 162.2 gm. However spoilage of nuts when stored for 60 days were 78.77 per cent in 9 months old nuts 31.55 per cent in 10 months old nuts, 13.10 per cent in 11 month old nuts and 7.62 per cent nuts in fully matured nuts. With the rich background on the pros and cons of coconut production discussed above, an attempt has been made hereunder to assess the coconut production practices in the study area.

Study Area Profile

State Profile: Tamil Nadu

Tamil Nadu is one of the spiritually, culturally, financially well developed states in India. The state lies between 8°5′ and 13°35′ in the North Latitude and between 76°15′ and 80°20′ in the East Longitude. The nearby states are Karnataka, Kerala, Andhra Pradesh and Union Territory of Pondicherry. The total geographical area of the state is 130058 sq.km. The total cultivated area and net area sown of the state is 6032718 and 5243839 hectares respectively. The normal rainfall in the state is 958.5 mm but the actual rainfall during the year 2005-2006 is 1304.1 mm, which is 36 percentage higher than the normal rainfall. The state has a population of 62.4 million as per the 2006 census. Out of total population, 55.96 and 44.04 percent of people belongs to rural and urban living category respectively. 73.5 percent of the population is literates. The density of the population is 480 per sq.km. The total work force in the state is 278.75 lakh, out of which 62.99 and 37.01 respectively. Nearly 85.22 percent of the working population occupies main worker category and 14.78 percent belongs to marginal worker category. (Special Commissioner and Director, Department of Economics and Statistics, Statistical Handbook of Tamil Nadu – 2006).

District Profile : Coimbatore

Coimbatore is one of the industrially developed and commercially vibrant districts in Tamilnadu. Well developed textile industry and other industrial base have earned a nick name of Manchester of South India. The district lies between 10 ^{0}10′ and 11 ^{0}30′ in the North Latitude and between 76^0 and 40′ and 77^0 30′ in the East Longitude. It is surrounded by the Nilgiris at the North.

District Profile : Tiru*ı *ur

Western Ghats at the West, Dindigul district at the South and Erode district at the East. It has three revenue divisions, nine taluks and ninteen blocks. The climate of the district varies widely. The Palghat Pass and Western Ghats considerably influenced the climate and rainfall. Normal rainfall in the district is 647.2 mm as against the state average of 925 mm (NABARD,2002). The predominant soil type is red and gravel with moderate areas of red loam, black loam and black clay soil.

The district has a population of 42.71 lakh accounting for 6.84 percent of the State's population as per the 2001 census. The rural population constitutes only 33.98 percent of the total population and the density of population is 566 per sq.km. as against the state average of 480 per sq.km. The total workforce of the district is 19.69 lakh against State's level of 278.78 lakh. Nearly 41.05 percent of the working population is directly engaged in primary sector activities, 30.05 percent in the secondary sector and the remaining 28.9 percent in the tertiary sector.

Total geographical area of coimbatore is 7.47 lakh ha. and account for 5.74 percent of the State's total area. Of the total geographical area, net sown area is 42.15 percent, 14.19 percent is land put to non-agricultural use, 7.74 percent fallow land and 1.85 percent is culturable waste, 0.89 percent is forest, 0.99 percent is Barren and cultivable land, 11.38 percent is permanent pastures and other grazing lands, 0.45 percent is miscellaneous tree groups and groves. The net irrigated area and gross irrigated area is 1.74 lakh ha and 1.85 lakh ha respectively.

Out of total area under crops irrigates, coconut occupies 56 percent maize 10.4 percent, fruits and vegetable 5.4 percent, sugarcane 4.9 percent paddy 4 percent and other crops 19.3 percent. Total rainfall during the year 2005-2006 is 863.1 mm as against the 814.8 mm of rainfall in the yester year. It accounts for 6 percent increase. The North East monsoon provides more rainfall to the district which contributes a major share of 58.6 percent, followed by South West monsoon (20.5 percent), summer shower (18.8 percent) and winter shower (2.1 percent). The gross cropped area during 2005-2006 stood at 333332 hectares as against 351425 hectares of pervious falsi year and showing a decrease to the tune of 5.4 percent. The district is leading producer of coconut in the state. Major other crops grown in the district includes paddy, maize, sugarcane, cotton, vegetable etc.,. (District Statistical Hand Book, Joint Director of Statistics, Coimbatore)

Division Profile: Pollachi

Pollachi division has a total geographical area of 11.66 sq.km. This division comprised of 4 blocks namely Pollachi north, Pollachi south, Anamalai and Kinathukadavu. There are 131 villages with the total population of 6.17 lakh with density of 458 per sq.km. Literacy level in this division is 61 percent.

This division has 197 registered factories with 50 percent workers of the population. There are 49 percent of agricultural labours of the total labour in this division. Of the total geographical area, the net sown area is 79,646 hectares and the irrigated area is 48,205 hectares. The major crops in this division are coconut, paddy, sugarcane, cholam, pulses, cotton and vegetables.

Block Profile: Anamalai

Anamalai block is a potential area with 3488 sq.km of total geographical area. The block is having net area of 40,676hectares sown at that 49.05 percent area is under irrigation (19,953 hectares). The major source of irrigation is through dry wells, borewells and tube wells and also through canal irrigation at Parambikulam Aliyar project.

The block comprised of 19 villages with 16,164 population. Literacy level is 57 percent. Agriculture is the major occupation in this block but also comprised of 48 registered factories, challenging the main occupation for its labour availability. Because 62 percent of the total population area workers in the factories. The major crops in this block are coconut, paddy, groundnut, pulses, cholam, cotton and sugarcane.

Demographic, Socio and Economic Profile of the Respondents

In this section demographic, socio and economic specifics such as sex, age, education, family size, family type, years of experience in agriculture, years of experience in coconut cultivation, land status, sources of irrigation, type of land, size of land, average annual income, sources of income, sources of borrowings, are analyzed with appropriate statistical tools for further understanding of the respondents.

Table 4.6
Demographic, Socio and Economic Profile of the Respondents

Profile	*Particulars*	*No. of Respondents*	*% of Respondents*
Gender	Male	266	86.9
	Female	40	13.1
Age	Young	50	16.3
	Middle	106	34.6
	Old	150	49.0
Education	Pre-matriculation	72	23.5
	Matriculation	126	41.2
	UG	54	17.6
	PG	22	7.2
	Illiterate	32	10.5
Family Size	Small	44	14.4
	Medium	214	69.9
	Big	48	15.7
Type of Family	Unitary	178	58.2
	Joint	128	41.8
No. of Years in Agricultural Business	Up to 15 yrs	108	35.3
	15-30 yrs	110	35.9
	Above 30 yrs	88	28.8
No. of Years in Coconut Cultivation	Up to 10 yrs	94	30.7
	11-20 yrs	124	40.5
	Above 20 yrs	88	28.8
Agricultural Land Status	Own	194	63.4
	Tenant	34	11.1
	Lease	78	25.5
Sources of Irrigation	Open well	162	52.9
	Bore well	108	35.3
	Canal	26	8.5
	Tank	8	2.6
	Lake	2	.7
Type of Land	Wet	200	65.4
	Garden	76	24.8
	Wet and Dry	24	7.8
	Wet and Garden	6	2.0
Land Holding	Small Farmer	206	67.3
	Medium Farmer	68	22.2
	Big Farmer	32	10.5
Income	Low (below 2 lakhs)	228	74.5
	Middle (2 to 3 lakhs)	38	12.4
	High above (3 lakhs)	40	13.1

Major Sources of Income	Coconut	8	90.8
	Business	18	5.9
	Others	10	3.3
Sources of Borrowings	Cooperatives	80	26.1
	Commercial banks	68	22.2
	Coconut Traders	104	34.0
	Finance Institutions	36	11.8
	Relatives	18	5.9

As to perusal of Table 4.6 that a very good majority of the respondents (86.9%) are males followed by females (13.1%). Indeed it is a common knowledge that men were found to be participating more actively than women in all business like wise agricultural business is also no exception. A sizeable portion (49 %) of the respondents falls in the category of old age group and majority of the farmers (41.20 %) have twelve years of formal education (matriculation).

Family size is classified into three for the purpose of analysis, namely small (up to 3 members), medium (3 to 4 members) and big (more than 4 members). It is quite appealing that majority of the coconut growers have medium size of family (69.9 per cent). Further the survey statement shows that 58.2 per cent of the coconut farming families belonged to nuclear type.

Experience helps the agrarian in judging climate, monsoon, season, and precautions to be undertaken in certain circumstances, market conditions etc. It is quite appreciable to note that considerable size of the respondents (35.9%) has 15-30 years of experience in agricultural business. Similarly a good majority of the respondents (40.5 percent) has 11-20 years of coconut cultivation. It shows that the young farmers are attracted towards coconut only after introduction of hybrid varieties.

Although there is a declining vogue towards agriculture profession, there is a good percentage of respondents (63.4 percent) possessing own status of their land. Now a days leasing is widely practiced in all sorts of business. Agriculture is also having no exception in this regard. Paucity of labour might have forced the coconut growers towards leasing. A significant proportion of the respondents (25.5 %) have leased out by the original owners. Tenant farming is also practiced by 11.1 percent of the respondents.

Water is considered as a life blood for all agricultural crops. The quantum of water availability decides the nature of crops to be cultivated. It is quite interesting to note that 52.9 percent of the respondents depend on the open well source. Nonetheless

35.3 percent of the respondents forced to use bore well.

Since the study area is highly fertile and having good source of irrigation, majority of the respondents (65.4%) possesses wet land followed by 24.8 percent of the respondents enjoy garden land. Nevertheless 7.8 percent of the respondents possess both wet and dry land. Further majority of the respondents (67.3 %) possess below 5 acres of land.

The survey results shows that a good majority of respondents (74.5 %) fall under the low income group followed by 12.4 and 13.1 percent of the farmer respondents covered under the middle and high income group respectively. Coconut occupies a major source of income to the majority of the farmer respondents (90.8 percent).

It is distressing factor to note that a considerable section of the respondents (34 %) credit requirement was met out by the coconut traders. No cumbersome process, no security, no legal formalities, immediate access might have motivated the respondents to avail credit from the traders. Nonetheless Cooperative banks (26.1 %) and Commercial banks (22.2 %) also actively participated in the lending operations.

Production Practices of the Coconut Growers

This segment deals with production practices followed by the sample respondents. To explore the production practices in detail a bundle of questions had been posed to the sample respondents like cropping pattern data such as reasons for coconut cultivation, variety of coconut cultivated, demand for male and female labour, area under important crops, type of input used; production and productivity data such as no of trees planted per acre, total possession of trees, age of the trees, distance between two trees, annual productivity per tree, production enhancement operations undertaken, rainfall level, availability of water, problems in input marketing, Is coconut cultivation is better than other crops, extension of coconut cultivation, dependency over coconut trees, labour mobility and source of sapling. The responses for the above issues are discussed hereunder:

Table 4.7

Reasons for Coconut Cultivation

S. No	*Reasons*	*No. of Respondents*	*Percentage*
1	Perennial	52	17.0
2	Profitable	118	38.6

3	Marginal labour force	134	43.8
4	Other	2	.7
	Total	306	100.0

It is evident from the table 4.7 that 43.8 percent of the respondents clearly viewed that marginal labour force is the motivating factor for the cultivation of coconut followed by 38.6 percent and 17 percent of the respondents expressed that profitability and perennial are the motivational factors for coconut cultivation respectively.

Table 4.8

Variety of Coconut tree Cultivation

S. No	*Variety*	*No. of Respondents*	*Percentage*
1	Grove crop	134	43.8
2	Peripheral crop	112	36.6
3	Both	60	19.6
	Total	306	100.0

The productivity and quality of coconut are directly linked with the variety of tree planted. For the past few years, there is enormous response for the grove varieties due to its productivity, years taken to yield, size, colour, taste etc. on the other hand peripheral variety is also more or less equally opted by the coconut growers due to the fact that life span of the tree, free from diseases, yielding years etc., It was verified in the survey that majority (43.8 percent) of the respondents preferred grove variety followed by 36.6 percent of the respondents opted peripheral variety. Nonetheless 19.6 percent of farmers opted for both the varieties to fetch the maximum benefits.

Table 4.9

Educational Level and Variety of Coconut Tree Cultivation

Educational Qualification	*Variety of Coconut tree Cultivated*						*TOTAL*	
	Grove crop		*Peripheral crop*		*Both*		*No.*	*%*
	No.	*%*	*No.*	*%*	*No.*	*%*		
Pre-matric	38	52.8	22	30.6	12	16.7	72	100.0
Matric	40	31.7	52	41.3	34	27.0	126	100.0
UG	40	74.1	6	11.1	8	14.8	54	100.0
PG	10	45.5	10	45.5	2	9.1	22	100.0
Illiterate	6	18.8	22	68.8	4	12.5	32	100.0
TOTAL	134	43.8	112	36.6	60	19.6	306	100.0

Chi-Square Test

	Value	df	Sig.
Chi-Square	48.482	8	**

** : Significant at 1% level

Chi square test was applied to find whether there is any significant relationship between educational qualification of the respondents and variety of coconut tree cultivation. The calculated value of Chi square is 48.482 which is greater than the table value of 20.090 at 1% level of significance which confirms that there is significance relationship between these two variables. *Hence hypothesis$_1$ is rejected.*

Table 4.10

Is Coconut Cultivation Better Than Other Crops?

Age in years	*Response*				*TOTAL*	
	Yes		*No*			
	No.	*%*	*No.*	*%*	*No.*	*%*
Young	40	80.0	10	20.0	50	100.0
Middle	92	86.8	14	13.2	106	100.0
Old	130	86.67	20	13.33	150	100.0
TOTAL	262	85.6	44	14.4	306	100.0

Chi-Square Test

	Value	df	Sig.
Chi-Square	2.422	3	Ns

NS: Significance at 5% level.

It is clear from the table 4.10 that a vast majority of the respondents irrespective of their age group (85.6 percent) opined that coconut cultivation is better than other crops. This may be due to the fact that profitability, perennial character etc. Whereas 14.4 percent of the respondents have negative opinion. This may be because of their bitter experience with the production and marketing aspects. The chi square value also confirms that (2.422) there is no significant relationship between the age group and their response about coconut cultivation.

Table 4.11

Are you Interested to Extend Area under Coconut?

Classification of the Respondents	*Response*				*TOTAL*	
	Yes		*No*			
	No.	*%*	*No.*	*%*	*No.*	*%*
Small	34	33.3	68	66.7	102	100.0
Medium	30	29.4	72	70.6	102	100.0
Big	66	64.7	36	35.3	102	100.0
TOTAL	130	42.5	176	57.5	306	100.0

Chi-Square Test

	Value	df	Sig.
Chi-Square	31.242	2	**

** : Significant at 1% level.

On the contrary to the findings of the previous table 4.10, majority of the respondents, irrespective of their classification (57.5 percent) were not interested to extend their area under coconut, which may be due to the fact that non-availability of vacant land, less water source, inability to manage the farm etc., and rest of the 42.5 percent of the respondents are interested to extend their area under coconut cultivation. The chi square value confirms that there is a significant relationship between the size of land holdings and their interest towards extending area under coconut.

Table 4.12

Dependency over Coconut Trees

Agricultural land status	*Dependency over coconut trees*				*TOTAL*	
	Alone farming		*Mixed farming*			
	No.	*%*	*No.*	*%*	*No.*	*%*
Own	140	72.2	54	27.8	194	100.0
Tenant	26	76.5	8	23.5	34	100.0
Lease	60	76.9	18	23.1	78	100.0
TOTAL	226	73.9	80	26.1	306	100.0

Chi-Square Test

	Value	df	Sig.
Chi-Square	788	2	Ns

NS: Significance at 5% level.

It is evident from the table 4.12 that a good majority of the respondents irrespective of their agricultural land status (73.9 percent) depends on only coconut farming. This may be due to the fact that to reap maximum benefit from the soil as well to enhance the quality of the products and to keep intact the quality of the soil for a longer period. The calculated value of chi square is .788 which is less than the table value of 5.991 at 5% level of significance which narrates that there is no significant relationship between agricultural land status and dependency over coconut trees.

Table 4.13

Demand for Male and Female labour

S. No.	*Demand*	*Male labour* *No. of Respondents*	*Percentage*	*Female labour* *No. of Respondents*	*Percentage*
1	High	66	21.6	38	12.4
2	Moderate	196	64.1	126	41.2
3	Low	40	13.1	102	33.3
4	Nil	4	1.3	40	13.1
	Total	306	100.0	306	100.0

Although coconut cultivation requires less labour force during the safeguarding period, it is very hard at present to search out both male and female labourers. This may be due to the fact that there is a less attraction, grave physical work moreover the wages is also not much than that of other work like construction. It is very perceptible from the table that there is a moderate demand both for male (64.1 percent) and female labour (41.2 percent). It was also confirmed with the result of informal discussion held with the cultivators that there is a heavy demand for semi-skilled workers like tree climbers and husk removers.

Table 4.14

Nature of Reaping

Average Annual Income (Rs)	*Nature of Reaping*						*TOTAL*	
	Coconut		*Copra*		*Tender Coconut*		*No.*	*%*
	No.	*%*	*No.*	*%*	*No.*	*%*		
Low (below 2 lakhs)	136	59.6	38	16.7	54	23.7	228	100.0
Middle (2 to 3 lakhs)	22	57.9	16	42.1	-	-	38	100.0
High above (3 lakhs)	30	75.0	10	25.0	-	-	40	100.0
Total	188	61.4	64	20.9	54	17.6	306	100.0

Chi-Square Test

	Value	df	Sig.
Chi-Square	30.312	4	**

** : Significant at 1% level.

Nature of reaping depends upon the variety of standing tree, market conditions, workforce availability, demand in the processing sector etc. From the table it is evident that a good majority of the respondents irrespective of their level of income (61.4 %) preferring to reap in the form of coconut rather than copra and tender coconut. The reason attributed for this state are hassle free harvest, permanent demand for coconut, profitability etc. It is obvious to note that almost all the low income category farmers are reaping in the form of tender coconut may be due to meet their unexpected expenses, regular flow of income to the family etc.,

Table 4.15

Type of Input used and Opinion about its Cost

S.No	*Type of Input*	*No. of Respondents*	*%*	*Opinion*	*No. of Respondents*	*%*
1	Organic	191	62.4	Costly	188	61.4
2	Inorganic	115	37.6	Low	118	38.6
	Total	306	100.0	Total	306	100.0

Organic farming is slowly gaining momentum among the farmers at present. It is apparent that majority of the eco-friendly sample respondents (62.4%) are following organic farming irrespective of its cost. Nonetheless 37.6 percent of the respondents still following inorganic farming owing to its cost effectiveness, readiness, easy availability, subsidy from the government etc.,

Table 4.16

Type of input Vs. Years of experience in agricultural business

Years of experience in agricultural business	*Type of Input used*				*TOTAL*	
	Organic		*Inorganic*		*No.*	*%*
	No.	*%*	*No.*	*%*		
Up to 20 yrs	66	61.1	42	38.9	108	100.0
21-30 yrs	60	54.5	50	45.5	110	100.0
Above 30 yrs	65	73.9	23	26.1	88	100.0
TOTAL	191	62.4	115	37.6	306	100.0

Chi-Square Test

	Value	df	Sig.
Chi-Square	7.899	2	*

* : Significant at 5% level.

An attempt was made to find significant relationship with two variables namely years of experience in agricultural business and type of input used. The calculated chi square value is 7.899 which is greater then the table value of 5.991 at 5% level of significance. So, it can be concluded that there is significant relationship between the variables.

Table 4.17

Source of Sapling

S. No.	*Source of sapling*	*Number of Respondents*							
		Small	*%*	*Medium*	*%*	*Big*	*%*	*Total*	*%*
1	Government	21	18.9	46	39.3	50	42.7	117	100
2	Private Agencies	76	43.2	52	29.5	48	27.3	176	100
3	Own seed	05	38.4	04	30.8	04	30.8	13	100
	Total	102	33.33	102	33.33	102	33.33	306	100

Quality of sapling decides the annual production and productivity of the coconut per tree. A good variety of sapling produces maximum benefits to the coconut grower. So, while selecting the source the farmers are very keen to cross check genuinity, origin, quality of sapling, variety etc. It is understood from the table 4.17 that a good number of farmers (176, 57.5%) opted the private agencies for their saplings followed by 117 (38.2%) chosen the Government departments. It is a universal awareness that private agencies are highly professional than that of Government sector. Moreover there is a pre and post sales services for the customers. It is appreciable that a very meagre number of farmers are using their own indigenous seed for their cultivation.

Table 4.18
Type of Trees Planted

Respondents Age	*Type of trees planted*						*TOTAL*	
	Ordinary		*Hybrid*		*Both*		*No.*	*%*
	No.	*%*	*No.*	*%*	*No.*	*%*		
Young	4	8.0	42	84.0	4	8.0	50	100.0
Middle	34	32.1	54	50.9	18	17.0	106	100.0
Old	84	56	40	26.67	26	17.33	150	100.0
TOTAL	122	39.9	136	44.4	48	15.7	306	100.0

Chi-Square Test

	Value	df	Sig.
Chi-Square	60.799	6	**

** : Significant at 1% level.

The young coconut peasant community would like to reap more benefits with in the available resources by way of using modern technology, hybrid coconut trees, drip irrigation etc., From the above table it is clear that a good majority (44.4 percent) of the respondents chosen hybrid varieties than that of ordinary variety in their cultivable land. It is interesting to point out that young age group respondents (84%) chosen the hybrid variety massively. It is further confirmed by the chi square test as the calculated chi square value is higher (60.799) than the table value, it is inferred that there is a significant relationship between the age of the respondents and the variety of the tree planted.

Table 4.19
Comparison of Periphery and Grove Variety in Terms of its Production and Productivity

S. No.	*Production particulars*	*N*	*Minimum*	*Maximum*	*Mean*	*Std. Deviation*
1	No. of trees planted per acre-Periphery	172	60.00	72.00	68.5116	2.8888
2	No. of trees planted per acre-Grove	194	60.00	75.00	68.9175	2.5170
3	Total possession of trees-Periphery	172	10.00	1200.00	271.0233	215.4883
4	Total possession of trees-Grove	194	30.00	4730.00	249.9897	492.3446
5	Age of the trees-Periphery	172	9.00	40.00	22.4767	6.5565
6	Age of the trees-Grove	194	4.00	40.00	10.7526	5.2660
7	Years taken to give yield-Periphery	172	5.00	9.00	6.9651	.5813
8	Years taken to give yield-Grove	194	4.00	7.00	4.2371	.7165
9	Distance between two trees-Periphery	172	22.00	28.00	26.6163	.7519
10	Distance between two trees-Grove	194	22.00	28.00	25.9794	.9328
11	Annual productivity per tree-Periphery	172	10.00	130.00	113.2093	17.6384
12	Annual productivity per tree-Grove	194	110.00	185.00	151.5258	19.3544

The table 4.19 distinguishes the grove and periphery tree in terms of number of trees planted per acre, age of the trees, years taken to give yield, distance between trees and productivity per acre. The minimum number of trees planted per acre is same for both the varieties i.e., 60 but the maximum has a meagre difference. The minimum age of the periphery and grove tree is 9 and 4 years and maximum for both the variety is 40 years. The mean age of the periphery tree is 22.4767 and standard deviation is 6.5565. The mean age of grove tree is 10.7526 and standard deviation is 5.2660. The minimum and maximum years taken to give yield by a periphery and grove tree is 5 and 9 years and 4 and 7 years respectively. The sample farmers in the study area follows uniform pattern of distance between the two trees. The mean annual productivity of the periphery and grove tree is 113 and 151 respectively.

Table 4.20
Mean scores of Production between Grove and Peripheral Crop

Variety of Coconut tree Cultivated	*Annual productivity per tree* Mean	S.D	No.
Grove crop	154.03	20.17	134
Peripheral crop	114.48	15.47	112
TOTAL	136.02	26.81	246

Chi-Square Test

	T	df	Sig.
Chi-Square	16.986	244	**

** : Significant at 1% level.

Table 4.21
Paired sample t-Test for Grove and Peripheral Tree Productivity

	Mean	*S.D*	*No.*
Annual productivity per tree-Periphery	110.83	21.04	60
Annual productivity per tree-Grove	145.93	16.19	60

	Paired Difference *Mean*	*Std. Deviation*	T	*Df*	*Sig.*
Annual productivity per tree-Periphery - Annual productivity per tree-Grove	35.1000	25.7226	10.570	59	**

** : Significant at 1% level

To know the significant difference between the grove and peripheral variety trees in the average productivity per tree t - test and paired sample t-test was applied by the researcher. The calculated t-test value is 6.986 which is greater than the table value of 2.596 at 1% level of significance. Since the calculated value is greater than the table value it is inferred that there is significant difference between these two varieties of tree in the average productivity per tree. The calculated paired sample t-test value is 10.570 which is higher than the table value of 2.662

at 1% level of significance. So, it is concluded that there is a paired difference among the two trees in the average of productivity per tree.

Table 4.22

Significant difference between grove and peripheral coconut and its total sales value per acre (in Rs.000s)

Variety of Coconut tree Cultivated	*Total coconut sales value per acre (Rs.000s)*		
	Mean	*S.D*	*No.*
Grove crop	38.03	6.72	134
Peripheral crop	29.94	4.97	112
Both	34.77	27.75	60
TOTAL	34.43	13.81	306

ANOVA

	Sum of Squares	*Df*	*Mean Square*	*F*	*Sig.*
Between Groups	3999.133	2	1999.567	11.181	**
Within Groups	54186.631	303	178.834		
Total	58185.764	305			

** : Significant at 1% level

The researcher applied one way ANOVA to find whether there is significant difference among the grove and peripheral coconut groups in the average total sales per acre. The ANOVA result shows that the calculated F-ratio value is 11.181 which is greater than the table value of 4.676 at 1% level of significance. Since the calculated value is higher than the table value it is concluded that there is significant difference among the variety of coconut groups in the average of total sales value per acre.

Table 4.23

Production Enhancement Operations Undertaken

S.No.	*Type of operations*	*No. of Respondents*	*Percentage*
1	None	12	3.9
2	Crop pruning	62	20.3
3	Rope support to flower bunches	92	30.1
4	Inter ploughing	98	32.0
5	Inter cropping	12	3.9
6	Others	30	9.8
	Total	306	100.0

Production enhancement operations such as crop pruning, rope support, inter ploughing, inter cropping gives definite result for the farmers in terms of productivity increase, product quality, life span enhancement etc., Table 5.23 confirms that in study area too the sample respondents are undertaking such kind of production enhancement operations for the purpose of availing utmost benefit.

Table 4.24

Credit Requirements of the Sample Respondents

Classification of the farmer	*None*		*Short-term*		*Medium-term*		*Long-term*		*TOTAL*	
	No.	*%*	*No.*	*%*	*No.*	*%*	*No.*	*%*	*No.*	*%*
Small	10	9.8	28	27.5	42	41.2	22	21.6	102	100.0
Medium	-	-	48	47.1	44	43.1	10	9.8	102	100.0
Big	2	2.0	26	25.5	56	54.9	18	17.6	102	100.0
TOTAL	12	3.9	102	33.3	142	46.4	50	16.3	306	100.0

Chi-Square Test

	Value	df	Sig.
Chi-Square	29.608	6	**

** : Significant at 1% level.

It is observed from the table 4.24 that majority (46.4 percent) of the respondents opined that they are badly in need of medium term loans for the purpose of meeting out the recurring expenses of the firm. It is appealing to note that 3.9 percent of the respondents have the financial capacity to look after the firm by their own wealth. The calculated value of chi square is 29.608 which is higher the table value of 16.812 at 1% level of significance which tells that there is significant relationship between these two variables.

Table 4.25

Are you purchase inputs from PACBs?

Classification of the Respondents	*Yes*		*No*		*TOTAL*	
	No.	*%*	*No.*	*%*	*No.*	*%*
Small	78	76.5	24	23.5	102	100.0
Medium	80	78.4	22	21.6	102	100.0
Big	83	81.4	19	18.6	102	100.0
TOTAL	241	78.8	65	21.2	306	100.0

Chi-Square Test

	Value	df	Sig.
Chi-Square	.742	2	Ns

Ns : Significant at 5% level.

Perusal of table 4.25 that a vast majority of the respondents (78.8%) have purchased inputs from PACBs. Indeed it is common knowledge that less price, good quality, subsidy from the government, nearness to the village, nominal overhead costs and no adulteration are the factors might have motivated the respondents towards the PACB. It is appealing to note that a good majority from the entire cross section of the farmers chosen the PACB as their source of input irrespective of their classification. Nonetheless a notable percentage (21.2) of the respondents were not opted the PACB for their input requirement due to the fact that advance money, supply of inputs for credit, non-availability in the PACB, farm gate supply from the non-institutional agencies. Chi square was applied to find the relationship between the purchase of inputs from PACB and classification of the farmer. The calculated chi square value is .742 which is lesser than the table value of 5.991 at 5% level of significance which prove that there is no significant relationship between the two variables.

Table 4.26

Reasons for Preferring PACB

S.No.	*Reasons*	*No. of Respondents*	*Percentage*
1	Good quality	64	26.6
2	Less price	77	32.0
3	Easy availability	70	29.0
4	No adulteration	30	12.4
	Total	241	100.0

It is evident from the table 4.26 that a considerable size of the respondents (32 percent) opined that less price is the prime motivating reason to prefer PACB for their input requirement followed by easy availability (29 percent) and good quality (26.6 percent). It is encouraging to view that 12.4 percent of the respondents preferred cooperatives for its quality supply of inputs.

Table 4.27

Average coconut production (value per acre) for Organic Input and Inorganic Input

Type of Input used	*Total coconut production cost per acre(Rs.000s)*		
	Mean	*S.D*	*No. of respondents*
Organic	35.40	16.47	191
Inorganic	32.83	7.35	115
TOTAL	34.43	13.81	306

t-test for Equality of Means

T	Df	Sig.
1.578	304	Ns

NS : Significant at 5% level.

It is evident from the table 4.27 that the mean score for organic input was 35.40 which is slightly higher than that of inorganic input score (32.83). In fact it is well known factor that the cost of production for organic input is quite higher than the inorganic input likewise the productivity is also equally high in organic input. The T-test was applied to find whether there is significant difference between organic input and inorganic input in the average coconut production cost per acre. The calculated t-test value is 1.578 which is less than the table value of 1.968 at 5% level of significance. Since the calculated value is less than the table value it is inferred that there is no significant difference between organic input and inorganic input in the average coconut production cost per acre. *Hence HO_3 : is accepted.*

Table 4.28

Average quantity produced per acre in alone and mixed farming

Dependency over coconut trees	*Total quantity produced per acre(No. 000s)*		
	Mean	*S.D*	*No. of respondents*
Single farming	8.97	3.95	226
Mixed farming	8.56	2.10	80
TOTAL	8.86	3.56	306

t-test for Equality of Means

T	Df	Sig
0.879	304	Ns

NS : Significant at 5% level.

It is understood from the table 4.28 that average coconut produced per acre through single farming is 8.97 which is slightly higher than that of mixed farming (8.56). The T-test was applied to find whether there is significant difference between single farming group and mixed farming group in the average quantity produced per acre. The calculated t-test value is 0.879 which is less than the table value of 1.968 at 5% level of significance. Since the calculated value is less than the table value it is inferred that there is no significant difference between single farming and mixed farming group in the average quantity produced per acre. *Hence HO_4 : is accepted.*

Table 4.29

Average quantity produced per acre by farmers

Classification of the respondents	*Total quantity produced per acre(No. 000s)*		
	Mean	*S.D*	*No. of respondents*
Small	9.32	1.92	102
Medium	9.35	5.42	102
Big	7.91	1.95	102
TOTAL	8.86	3.56	306

ANOVA for Total quantity produced per acre (000s)

	Sum of Squares	*Df*	*Mean Square*	*F*	*Sig.*
Between Groups	138.927	2	69.464	5.651	**
Within Groups	3724.348	303	12.292		
Total	3863.276	305			

** : Significant at 1% level

It is clear from the table 4.29 that average production per acre is high among medium farmers (9.35) than small (9.32) and big farmers (7.91). One way ANOVA was applied to find whether there is significant difference among the type of farmer groups in the average quantity produced per acre. The ANOVA result shows that the calculated F ratio value is 5.651 which is greater than the table value of 4.676 at 1% level of significance. Since the calculated value is greater than the table value it is inferred that there is significant difference among the type of farmer groups in the average quantity produced per acre. *Hence HO_5 : is rejected.*

Sum Up

Production practices are mostly characterized by the farming culture, environment, local conditions, economic and social conditions of the coconut growers. Since, the study area is coconut concentrated, most of the coconut growers are practicing the same method coconut of cultivation. But the efficiency differs among the cultivators on the basis of agricultural experience, soil condition, availability of water, technology, awareness, organic and inorganic input usage etc.

REFERENCES

1. Anirudha Sharma, Villagers can do well with better Marketing, *Indian Farmer Times*, Vol. 10, No.11, 1993.
2. Aravindakshan, M., India attains the premier position in coconut production, *Indian Coconut Journal, Kochi*, CDB, Vol. 28, No.3, 1997.
3. Bhatia, B.M and others, "Food, Famine and Population", Population in India's Development, 1974.
4. Coconut Development Board, Coconut Situation in India, Kochi, 1997.
5. Food and Agricultural Organisation (FAO), Marketing of Agricultural Products, Rome, 1960, p.16.
6. George, M.V., Vijaya Kumar and others. Trend in Area, Production and Productivity of coconut in India, *Indian Coconut Journal*, Vol. XXII, No. 647, Kochi, 1991.
7. Haridoss and Chandran, Marketing systems, costs, margins, price spread and marketing problems of coconut—A case study of coconut growers and traders in Tamil Nadu, *Indian Coconut Journal*, Vol XXVII, Nov. 1, May 1996.
8. Hough, E.M., *The Cooperative Movement in India*, Oxford University Press, Calcutta, 1966.
9. Jnanadevan R and Prakash R., An analysis of constraints in increasing coconut production in Kerala, *Indian Coconut Journal*, No.4 Auguest 1993.
10. Markose, V.T.. Coconut Development in the last decade, *Indian Coconut Journal*, Vol.25, No.7, CDB, Kochi, 1994, P. 12.
11. Memoria, C.B.,. *Agricultural Problems of India*, Kitab Mahal, Allahabad, 1998.
12. Rathinam, R. and others, Global scenario of coconut oil, Paper presented at OFIC Kula Lumpur, Malaysia, October 2002.
13. Ron Harris, The Coconut Industry- Utilization and Marketing , *The Planter*, Kuala Lumpur, Malaysia 1992.

14. Satyasunaram, I. Market survey—Coconut Palm: Diversification holds the key, *Facts for you*, New Delhi: Indian Book House, Vol. 18, No. 5, September 1997.
15. Subburaj, B., Coconut Processing industry in Tamil Nadu, with special reference to Kangayam, Paper presented in the National Meet on Coconut, held at Kochi, November, 1996.
16. Subburaj.B., Marketing of Coconuts—Practices and Problems in Tamil Nadu- A Research Monograph, Gandhigram Rural Institute May 2000.
17. Thampan, P.K., *Global Competitiveness of Coconut Industry*, Kochi: Pekay Treecrops, Kochi, 1995.
18. Thampan, P.K., Glimpses of Coconut Industry in India, CDB, Kochi 1988.
19. Venkatachalam V., Investment opportunities in coconut in Tamil Nadu, *Indian Coconut Journal*, 1997, June, p. 8.

5

Marketing of Coconuts: Practices, Problems and Prospects

Introduction

Production and productivity of agricultural products are very much linked with marketing of the same in the raw as well as processed form. When there is an efficient marketing system, farmers are assured of reasonable or attractive prices for their produce and wastage of produce and expenses in different stages of marketing are minimized. Increase in net income for the produce will encourage farmers to bring more area under particular crop, to invest in more inputs and to adopt latest agricultural technology. This will lead to improvement in production and productivity, which in turn will improve the national economy. The surplus production can be utilized for export and also the manufacture of diversified and value added products. It will also help to increase employment opportunities. Increase in productivity helps to reduce unit production cost of the produce and make sale rate of the produce more competitive even after assuring very reasonable or attractive remuneration to the products. The customers will also be satisfied by getting the produces at reasonable rates. The competitive rate will help export promotion and earn lucrative foreign exchange. Thus production, productivity and marketing are very much linked and poor efficiency or weakness in any of these in respect of agricultural produce will affect future development and prospect of the crop. This situation is very relevant in the case of coconut and its products like coconut oil, which face stiff competition from other countries and from other cheaper substitute oils even with in the country. Unfortunately in the past, in the independent India, due emphasis was not given to marketing aspects of the agricultural produce. Many schemes were formulated and implemented towards the goal of achieving higher crop production. Due to lack of planning with regard to marketing, the farmers were deprived of reasonable remuneration and it affected productivity as well as desired level of further development of many crops.

Crisis experienced by cocoa cultivators in Kerala about a decade ago due to sudden withdrawal of multinational company, which was enjoying almost monopoly in cocoa purchase, is a good example for such a situation. Fortunately the crisis did not last long because of the timely intervention of the Kerala State Cooperative Marketing Federation (MARKFTFED) to procure and do primary processing, such a procurement operations helped to save farmers from crisis. This example reveals that cooperative sector can alone rescue farmers from such crisis by extending its activities even to the interior rural areas with its wide spread procurement network, infrastructure and expertise. The above example of agricultural products stresses the need for a long term vision in marketing and the importance of cooperative organisation like marketing federation.

India has been evincing good track record of performance in Agriculture. Since independence agricultural production and productivity have increased manifold. India attained self-sufficiency not only in food production but also in production of other agricultural commodities. As for coconut, India is one of the major coconut growing countries in the world. It occupies second position in terms of nuts production among 16 countries producing in the world (APCC 2006). It is estimated that coconut is grown in 1.93 million hectares and 12, 832 million nuts are produced annually. Among coconut growing states in India, Tamil Nadu stands third in the area under cultivation and production of coconuts (Directorate of Economics & Statistics, GOI, 2006). Coconut is grown in 30 districts of Tamil Nadu. There has been a spectacular increase in the area and production of coconuts in the state during the last five decades. The area, which was only 0.66 lakh hectares in 1951, has increased to 110 lakh hectares in 1976 and further went up to 370 lakh hectares in 2006. Moreover the productivity of coconut is around 13,133 nuts per hectare, while it is only around 7046 nuts in Kerala (CDB, 2006). It is estimated that Tamil Nadu will produce a marketable surplus of about 4000 million nuts which could meet not less than 30 percent of India's demand for coconuts for both edible and non-edible end use by the turn of this century (Subburaj 1996).

In India, the process adopted for the marketing of coconut differs from region to region. A recent publication of the Coconut Development Board, "30 Years of Coconut Industry" explains the consumption pattern of coconut in the country. Out of the total production of coconut in the country, 35-40 percent is used for value addition, mainly copra making and the remaining as direct consumption either as a food, drink, and beverage. In Kerala around 41.2 % is consumed as raw form, while 58.5 % is used for making copra. Tamil Nadu also utilises around 47 % of the production for milling copra as well as desiccated coconut and around 10 % for tender nuts, the remaining for raw consumption and traded outside the state, mostly in Northern and Western India (Subburaj B. 1998). It is estimated that around 25% to 40% of the production in Karnataka and Andhra Pradesh is consumed in the form of tender nuts and the rest for manufacture of ball copra, besides the use in raw form. The practice followed is that the fully matured raw coconuts are kept in specially designed atties for eleven to twelve months. When the water inside is fully absorbed, the nuts are dehusked and the kernel detached by breaking the shell and the round/oval shaped kernel obtained is called the edible ball copra. There is a good demand for this product in the up-country markets like Cuttack, Jaipur, Agra, Ahammedbad, Delhi, Bombay etc.

In Orissa, about 60% of the production is marketed as tender coconuts (Sugantha Gosh 1997) while 80 % in West Bengal. In the case of the Andaman & Nicobar Islands, which is a traditional belt for coconut, the consumption pattern differs in between the islands. It is estimated that about 42% of the coconut produced in the islands are consumed for edible purposes, 30% for milling copra and 28% for tender coconuts. (Retinam, Thomas and Rajeev 2001)

It is widely claimed that in all the years, farmers have been deprived of the benefits of increased production in monetary terms. Equalization of demand and supply is far from anybody's guess. Since farmers can realize the advantages of increased production/productivity only through better marketing of their produce, marketing therefore is as important as production (Memoria, C.B. 1995). To a coconut farmer, marketing is something more than production, due to certain inherent features that neither the production can be controlled nor regulated in tune to market changes due to predominance of natural forces affecting the production functions nor can marketing be performed in tandems to market requirements due to obvious reasons, particularly the limiting factors governing the farming community in general and the coconut farmers in particulars. The coconut farmers remain with chronic problem. It is apt to recall the words of Dantwala Committee of Cooperative Marketing (1966) and, Food and Agricultural Organisation (1988) that agricultural production is abundant but beset with problems in distribution. The distribution system should be terse, free from imperfections so that it can provide remunerative and assured price to farmers. It should benefit both the producers and consumers by delivering maximum satisfaction to one and all, involved in the marketing process. Creating 'perfected marketing system' therefore is a 'must'.

This chapter discloses the strategies employed for disposal of coconuts by farmers, pattern of disposal, mode of disposal, institutional and non-institutional functionaries an their procurement practices, sale practice, opinion about the price offered by the market functionaries, channel member choice and selection criteria of farmers and problems in the disposal of coconuts at farm level etc., are discussed. It concludes by offering suggestions of farmers for effective disposal of coconuts at the level of primary markets.

Current System of Marketing of Coconut and its Products

The current system of marketing of coconut and its products is, by and large, unscientific and unorganised and is almost perpendicular integration. Although coconut is made a notified commodity in most of the states, the marketing lacks systematic organisational setup. In the absence of this, the exploitation by middleman and other functionaries is rampant. In almost all the primary markets, the middlemen normally utter the price. In addition to this, various kinds of malpractices are also commonly seen in most of the markets. In India, almost all the coconut growers are small or marginal holders so far not yet brought under the organized set up. Hence they are forced to sell their produce to the channel intermediaries because of contingent pressures. The private traders advance money to the growers and exploit them in every nook and corner. For instance, in Arsikere market in Karnataka, 110 coconuts are accounted for only 100 coconuts while buying from the growers. Apart from this these private middle functionaries openly practice deductions in weight and various kinds of commission and omissions.

Like the fresh coconuts, private traders also mostly control the marketing of copra. Here too these are private merchants take advantage of the falling price situation of coconut oil and poor economic status of the coconut growers.

Regarding the marketing of coconut oil, leading oil millers have a systematic distribution channel for their oil, either directly or through middle agencies. The marketing of oil is not a problem as such, but the preference of other edible oil to the coconut oil eroding demand and erratic prices of coconut oil are bottlenecks hindering the smooth marketing of coconut oil.

Chronological perspectives of disposal of coconuts

It was observed that some 100 years back farmers did not face problems in the disposal of coconuts. It became a problem to them only 40-50 years from now, when coconut trees are considered as an important source of income. The increased production of coconuts required the farmers to bring their surplus coconuts (i.e., the residue after local sale) to nearby local market places and weekly marketing 'shandies'. These apart, itinerant traders entered the village and procured dehusked coconuts from farmers. In due course, farmers felt the importance of coconut traders coming into villages. However, farmers with coconut groves in particular, found it difficult to dispose of their entire surplus coconut in the local / daily markets. Therefore, as an effective means of disposal of coconuts, they adopted the practice of leasing coconut trees, besides selling coconuts to traders. All the ways and means of disposal of coconuts adopted by farmers came to stay in villages, as problems related to marketing of coconuts became complex.

Today, farmers depend highly on traders for disposal of their coconuts. Traders play a significant role and form part and parcel of the system of coconut marketing in villages.

Table 5.1

Chronological view on the disposal of coconuts by farmers

Details	*Farmers under different categories*		
	Big	*Medium*	*Small*
Some 100 years before			
Mode of disposal of coconuts	Sale to neighbours and in local sandy	Sale to neighbours	Sale to neighbours
Dependence of traders	Nil	Nil	Nil
Some 40-50 years before			

Mode of disposal of coconuts	Local sale to traders and neigh-bours, sale at the local market	Sale at the local market and neigh-bours, and a little to traders	Sale to neighbours and in the local market
Dependence of traders	High	High	Low
At Present			
Mode of disposal of coconuts	Farm gate sale to traders, Comission Agents and Industrial User	Farm gate sale to trader, Commission Agents and Small Scale Industrial user	Sale in the local market, traders and oil crushing mills
Dependence of traders	High	High	High

(Subburaj B., 2004).

Marketing Practices of Coconut among the Respondents

In agricultural marketing there is no uniform practices followed either by the producer or marketer. It is purely depends upon the environment in which the production and marketing takes place. Like wise in coconut marketing also there is no uniform marketing practices followed through out India. In study area the sample coconut growers follows their own marketing practices to market their produces. Marketing practices refers the pattern of sale of coconut, mode of disposal of coconut, sale practices of sample respondents and their choice of market functionaries. The marketing practices followed by the sample respondents in the study area are discussed hereunder:

Pattern of Sale of Coconut

Pattern of sale can be done in to two type viz., selling of coconut with husk and without husk. Husk is used as a major raw material for the production of many byproducts like coir mat, coir rope, toys etc., Selling of coconut in either one of the pattern have its own impact on the profitability and income rate of coconut farmers. Most of the small and medium farmer's financial and infrastructure position doesn't allow them to undertake value addition to their produces. The following table shows the pattern of sale of coconut undertaken by the different category of the respondents.

Table 5.2
Pattern of Sale of Coconut

S. No.	*Pattern of sale*	*Category of the farmers* *Small farmer*	*Medium farmer*	*Big farmer*	*Total number of respondents*
1	With husk	101 (37.2%)	98 (36 %)	73 (26.8%)	272 (100%)
2	Without husk	01 (2.9%)	04 (11.8%)	29 (85.3%)	34 (100%)
	Total	102	102	102	306

It is well understood from the Table 5.2 that a very good number of the respondents (272) disposed their coconut with husk. Within this category a sizable portion of the place was occupied by the small and medium category farmers. This kind of value addition is not commonly practiced among the small and medium farmers due to the fact that paucity of manpower, less financial capacity, lack of infrastructural facility, avoidance of time delay in selling, minimizing the chance of price reduction, poor knowledge on technology etc., and rest of the (34) respondents are capitalizing and utilizing the benefits of husk by way of disposing their coconut in the form of without husk which is mostly practiced by the big farmers.

Disposal Practices of Coconut Farmers

Generally the right mode of disposal of coconut fetches a good price to the produce of coconut growers. In the study area the farmers analyze the pros and corns of each and every mode before their disposal of coconut. Inspite of the advantages in some mode, the respondents prefers their own mode because of various reasons like financial support, paucity of time to look after farm, migration to town, subsidiary occupation and some preoccupied business. Based on disposal practice, the coconut growers, further categorized into four namely leaser farmers, opportune farmers, mixed practice farmers and debt bonded farmers. An in depth analysis in this regard is furnished below:

Table 5.3
Disposal Practices

Category of the farmer	Disposal practices								TOTAL	
	Leaser farmer		Opportune farmer		Mixed practice		Debt-bonded farmer		No.	%
	No.	%	No.	%	No.	%	No.	%		
Small	18	17.6	42	41.2	16	15.7	26	25.5	102	100.0
Medium	34	33.3	42	41.2	20	19.6	6	5.9	102	100.0
Big	16	15.7	68	66.7	18	17.6	-	-	102	100.0
TOTAL	68	22.2	152	49.7	54	17.6	32	10.5	306	100.0

Chi-Square Test

	Value	df	Sig.
Chi-Square	52.677	6	**

** : Significant at 1% level.

It is clearly evident from the table 5.3 that considerable percentages of the respondents (49.7 %) are falling under the category of opportune farmers which provides ample freedom to follow their own disposal strategy. On the other hand those who are opted for leasing (22.2 %) and debt-bonding (10.5%) don't have a right to decide the disposal strategy. Nevertheless 17.6 percent of the respondents are following mixed practice of disposal for the purpose of capitalizing the market situation. Further scrutiny of the table shows that a sizable portion of the small and medium farmers are under the clutches of leasing and debt –boding system due to the fact that their less financial holding capacity on the contrary a good majority of the big farmers are dictating the market by way of following opportune system.

The chi square test was employed to know the significant relationship between the category of farmers and their disposal strategy. The calculated chi square value is 52.677, which is greater than the table value of 16.812 at 1% level of significance, so it can be concluded as there is a significant relationship between the variables.

Table 5.4
Disposal Practice Vs Years of experience in Coconut Cultivation

Years of Experience in Coconut Cultivation	*Leaser farmer*		*Opportune farmer*		*Mixed practice*		*Debt-bonded farmer*		*TOTAL*	
	No.	*%*	*No.*	*%*	*No.*	*%*	*No.*	*%*	*No.*	*%*
Up to 10 yrs	28	29.8	40	42.6	16	17.0	10	10.6	94	100.0
11-20 yrs	28	22.6	64	51.6	20	16.1	12	9.7	124	100.0
Above 20 yrs	12	13.6	48	54.5	18	20.5	10	11.4	88	100.0
TOTAL	68	22.2	152	49.7	54	17.6	32	10.5	306	100.0

Chi-Square Test

	Value	df	Sig.
Chi-Square	7.541	6	Ns

Ns : Significant at 5% level.

The researcher desires to correlate the two variables viz., disposal practice of the farmers and years of experience in coconut cultivation to find out whether there is any significant relationship between the variables. The calculated chi square value is 7.541 which is less than the table value of 12.592 at 5% level of significance which narrates that there is no significant relationship between these two variables.

Leaser Farmers

Leaser farmer is one who lease out his trees to somebody for a lease period of one or two years. Generally this lease agreement is done with or without formal registration. Under this arrangement and contract, the coconut growers loose their right to decide the disposal strategy irrespective of the market conditions and profitability.

Table 5.5
Reasons for Leasing

S. No	*Reasons for leasing*	*No. of Respondents*	*Percentage*
1	Redemption of prior debt	6	8.8
2	To meet the marriage expense	20	29.4
3	Old age	6	8.8
4	Migration to town	36	52.9
	Total	68	100.0

It is understood from the table 5.5 that most of the respondents leased out their trees due to the financial constraints and other personal reasons like old age and migration to town and other places. Among the reasons for leasing migration occupies a major share (52.9%). A considerable portion of the respondents (29.4%) leased out their trees for the purpose of to meet the marriage expenses of their children and family members. It is a tradition in India, that the marriages are conducted for the sake of exposing their wealth and to enhance their image among the public. This superfluous practice leads the farmers to end up with more financial burden followed by leasing out coconut trees. Further probe on the reasons for leasing out the trees was made through Focus Group Discussion (FGD) among a group of farmers. They said:

> Although our occupation is restricted in to coconut cultivation, we do not want our son's and daughter's to be fully engaged in coconut cultivation, rather they should be fed with high quality education, enjoy better standard of living, access to better infrastructure, access to high-tech medical facilities. Hence, migration to town and leasing out the trees

(FGD held among coconut growers of Jallipatti village, Anaimalai Block)

Opportune Farmers

Opportune farmers are independent farmers, who have not availed themselves of financial assistance either as loan or as advance from any of the channel members including coconut processing centers and individuals by leasing coconut trees or debt bondage or both, and those who can sell their coconuts to whoever they wish. But in practice it can be done only those who have financial strength and freedom.

Benefits of Being an Opportune Farmer

The result of FGD in this regard is discussed as follows:

The production operations and harvesting coconuts are made easy to opportune farmers as labour force is supplied by traders. Further, opportune farmers do not face problems in negotiating wages for labourers, including perks. Since coconut tree climbers are assured continuous employment from traders and agents, they adhere to a set wage structure. To avoid making excess payment, coconut farmers prefer to employ labourers with the help of traders / buyers. Further, the post-harvest activities such as sorting, counting, and loading of coconuts are carried out by tree climbers when they are engaged with the help of traders. The opportune farmers sell their coconuts as per their choice. They sell their coconuts on cash at the market price.

Mixed Practice Farmers

Farmers who lease a certain number of coconut trees and retaining some for their own harvesting are referred as mixed practice coconut farmers. This system of mixed practice prevails due to the fact that shortage of manpower, differences in age and productivity of the coconut tree, location, fertility of land, sources of irrigation, facilities for transportation, cropping pattern, distance from market centre / urban areas, local demand for tender coconuts etc. Further, the nature of ownership over the coconut farm is also attributed as one of the reasons. However, differences in the age and productivity of coconut trees distance from the market centre / urban area and the demand for tender coconuts are found to be the major reasons in the study area.

Debt Bonded Coconut Farmers

Debt bonded coconut farmers are those who sell their coconuts to buyer-cum-creditors till the loan amount and the interest is repaid. Such types of farmers are prevalent in small and medium farmers' category and not common except in villages where coconut trees under dry farming are found.

Results of Focus Group Discussion

The outcome of key informant interview, observation and FGD are discussed hereunder:

Distribution Channels

Farmers dispose of their coconuts through two channels namely direct and indirect. The direct channel is very simple wherein coconut farmers sell their products directly to consumers. In indirect channel, there are a number of intermediaries involved. A detailed analysis in this regard is furnished hereunder:

Direct Channel

Direct channel is popular among the farmers who are possessing a very few number of coconut trees, free from debt bondage and coconut farms near the market centers have adopted this channel. However, this channel is not found very prominent in the study area.

Indirect Channel

The most commo ι and prominent channel for distribution of coconuts by farmers is indirect channel. Two-thirds of coconut farmers have adopted this channel in the study area. Depending on the number of channel intermediaries involved, the indirect channel can be classified into (i) single layer channel, (ii) double layer channel, (iii) triple layer channel and (iv) four layer channel. The following diagram exhibits channel route in the indirect channel.

Indirect Channel Layers

Channel I: Farmer→Retailer→Consumers

Channel II: Farmer→Trader→Retailer →Consumers

Channel III: Farmer→Trader→Wholesaler→Retailer →Consumers

Channel IV: Farmer→Commission Agent →Wholesaler→Retailer→ Consumer

Channel Functionaries

In the indirect channel, the prominent channel intermediaries include traders and their agents, wholesalers and their agents, processing units and their agents and retailers. Since cooperative marketing bodies as well as regulated markets do undertake agent functions on behalf of farmers, they are referred to as farmers' agents. All the channel members including agents perform a number of functions in the process of movement of coconuts as well as coconut products from farmers' field to the doorsteps of consumers. Hence they are called channel functionaries.

Traders

Among all the channel intermediaries, traders can play an important role. Based on nature and type of activities undertaken by the traders in the study area, there are three categories of traders namely itinerant traders, independent traders and debt bonded traders. Each of them has their own uniqueness and peculiarities therefore, their contribution to the channel system is significant.

Itinerant Traders

They are local traders, buying in small quantities directly

from coconut farmers for direct sale to consumers. They procure mostly de-husked coconuts directly from marginal and small farmers in villages and local market places on ready cash. They do not employ brokers / agents. They effect sales mostly to ultimate consumers by avoiding unnecessary channel intermediaries which enables them to offer better price to farmers as well as to earn good profit out of the business.

Preference for Itinerant Traders

Coconut farmers who possess small number of trees, who are free from debt bondage and who would like to sell their coconuts in the village itself, prefer itinerant traders coming to the village, due to the following advantages:

- While negotiating with traders over price, farmers have every probability to quote the price prevailing in local retail/ petty shops. Thus, they have elasticity in price determination.
- There is no loss due to quality and quantity discounts.
- The time and cost involved in taking the coconuts for sale in the local market and other incidental expenses are avoided.
- No brokerage/commission is paid as no such agents are involved/entertained.

However, there is a disadvantage that more than one push cart load of coconuts cannot be sold as the itinerant traders do not buy more than that at one time. They operate on a small scale and that too, in local markets.

Sale in Local Market

Coconut farmers possessing coconuts in small quantity that are almost discarded by itinerant traders in villages and those farmers who do not like to dispose of their coconuts through agents / independent traders, sell their coconuts in local market places. During flush season, farmers used to resort local market place rather than village trader for the disposal of coconut to fetch high price. The buyers of coconut in the local market place include ultimate consumers, resellers and agents. Ultimate consumers include individual consumers and institutional consumers such as hotels, hostels of educational institutions, government departments and those who use coconuts for producing other products in cottage and village industrial scale. Resellers in the local market places include itinerant traders and retailers. In the market place, eight out of every ten buyers are found to be itinerant traders. While selling to ultimate consumers, farmers generally sells quality coconuts unless consumer pesters them.

Independent Traders

They are local traders buying coconuts in large quantities either directly from farmers or through agents / brokers on cash; for direct sale to wholesalers or processors or industrial users or retailers. They are not agents of wholesalers, processors or industrial users of coconuts and they do not have debt bondage with their buyers. Depending on the mode of sale, big farmer-cum-traders and lease-holder-cum traders can also be identified as independent traders. The majority of independent traders operate through their brokers and agents. Many such traders are found in villages fall under High and Moderate category. Existence of more of peripheral trees, low productivity of coconut trees due to scanty production enhancement operations, wide-seasonal variations or/and unstable water resources for irrigation, high cost of harvest and assembling of coconuts are the major reasons for small number of independent traders in Low category villages. Village wise analysis also indicates prominence of trader-cum-lessees in High category villages, local farmer-cum-traders in Low category villages and independent traders with their agents in Moderate category villages.

Functions of Independent Traders

Independent traders occupy a unique place in the channels of distribution of coconuts as they perform the following functions.

- Assembling of coconuts from farmers: By supplying labour force, transportation facilities, they enable farmers to assemble coconuts at the farms.
- Sorting and grading coconuts: They grade and sort coconuts according to the size and kernel maturity depending on market requirements.
- Credit requirements: Most of the short and long term credit requirements of coconut farmers are being met out by the traders either on lease of debt bondage over coconut trees or on personal surety. In fact, traders act as lenders of the last resort.
- Management of seasonal fluctuations: By building up large inventory of coconuts during flush days and

releasing them during lean days, they exercise control over the demand for and supply of coconuts.

- Regular supply: Industrial users are served with required quantity and quality of coconuts throughout the year.
- By products distribution: Since coconut husks and, in a few cases, coconut shell is available in large quantity, the coconut fiber units and paper mills are benefited by them.
- Other marketing functions: Traders undertake most of the Concentration functions like assembling, transportation and storage, standardization and grading, a little of processing activities etc. and some of the Dispersion functions like sorting and packing of coconuts, provision of marketing finance, risk bearing etc. They also play an important role by performing the most important market Equalization functions by exerting influence over demand and supply forces by maintaining adequate inventory, especially at secondary market places. Besides, they have been promoting marketing channels by allowing quality as well as quantity discounts, credit sale facility, and market intelligence. Thus, traders occupy a unique place in promoting channels for distribution of coconuts.

Procurement Practices of Traders

To enable procurement of coconuts in villages, traders maintain a band of tree climbers, agents, and brokers. Agents and brokers are paid commission according to their contribution to trade, including supply of information about the time and quantum of availability of coconuts in villages. Thus tree climbers are not only acting as canvassing agents of traders but also as link agents between traders and farmers.

Farmers said that traders have been preparing a schedule depicting the location and ownership of the farm, number of trees and season wise availability of probable quantity of coconuts from farmers. There are instances where traders have hesitated to harvest coconuts when farmers approached them a few days before the due date of harvest. On the other hand, traders have sent their tree climbers to the coconut farm where harvest was not done within the due date. Thus, by maintaining close watch and contact with farmers, traders procure coconuts. Further, traders make the job of harvesting and allied post-harvest functions to be done at the farm itself easier for coconut farmers, and hence farmers prefer traders.

Allowances

However, coconut farmers do not feel happy over some practices of traders. It is reported that traders insist on farmers to providing various allowances while procuring coconuts 5 to 10 for every 100 coconuts as quantity allowance, another 2 to 6 for every 100 coconuts as quality allowance and yet another 10 to 15 coconuts as transport allowance. Further, with a view to obtaining advantageous price for coconuts, traders deliberately grade coconuts with the help of their tree climbers who are very close to them. During final counting of coconuts for computing the sale proceeds to be payable to farmers, coconuts ranging from 10 to 15, which are left over, are also taken away by traders, for which no price is paid. Further traders used to fix low price even for first grade coconuts which is lower than the price at which traders sell coconuts.

Although traders agree to pay the entire sale proceeds in one lump sum immediately after the harvest, in practice, they settle the payment only after 7-15 days, that too, in most cases, in two installments. As traders are few in number, they join hands in fixing the marked price of coconuts. Based on the marked price, they decide the price to be administered among farmers through their eloquent negotiations. Despite all these discrepancies, farmers prefer traders as their channel members.

Debt Bonded Traders

Existence of debt-bonded traders is found especially among the small farmer category. They are mostly akin to independent traders except in their sale practices. Independent traders have options to choose their buyers as they do not have any tie-up arrangements with their reseller-buyers (wholesalers). Depending on the credit worthiness and personal background of the debt bonded traders, fresh loans are provided then and there by their wholesale traders / industrial users. Debt bonded traders bring coconuts at their own risk and cost to the premises of creditor-wholesalers. Coconuts are graded and sorted. Debt bonded traders have no hold over their coconuts. Usually the price at which they sell coconuts to their masters is 0.50 paise to Re. one per nut lower than the price offered to independent traders. Further, they have to provide 5-10 coconuts for every 100 coconuts for every 100 coconuts as quality allowances to wholesalers even though the coconuts are already graded twice, once in the place of harvest and next, in the premises of the wholesaler. Further, not less than 75 per cent of their sale proceeds are adjusted towards their loan accounts. They are provided fresh loans to take up procurement operations further. Thus, the debt bondage Marketing Practices and Channel Members is kept alive.

Wholesalers

Among the channels of distribution of coconuts, wholesalers occupy an important place. They are found in regional and mandi markets. They are the de-facto kingpins in the coconut marketing system. As channel functionaries, they have been doing a number of marketing functions, and, as channel promoters, their role cannot be underestimated. According to the observation made in the field, two types of wholesalers are found, selling within Tamil Nadu and selling to other states. They also act as agents of large scale coconut processing / confectionery industries located within and outside Tamil Nadu. They assemble coconuts at mandi/ secondary market level in large scale and stock them and exercise control over demand for and supply of coconuts to their advantage. They also have their offices in places where production of coconuts is high and where the demand for coconuts exists. Within Tamil Nadu, most of them sell to their customary retailers and industrial users of coconuts. Only a few of them sell to wholesalers located in inter-state market centers.

Functions of Wholesalers

No wholesaler has embarked upon export of coconuts. Given below are a few of their major functions.

- Assembling of coconuts in large scale at secondary market level.
- Processing of coconuts according to the market requirements – dehusking, partial dehusking, grading and standardization. However, most of them do not undertake copra making or activities related to product diversification.
- Distribution of coconuts in local and distant markets.
- Supply of coconuts to industrial users, large scale processors like oil mills and established confectionary manufacturing industries regularly.
- They extend financial assistance to both traders and farmers, as also to buyers of coconuts including retailers.

If they had not extended credit sale facilities, the entire channel at all levels could have collapsed.

- They promote the channels by providing both pecuniary and non-pecuniary benefits to those who constitute their distribution network. However, coconut farmers are dissatisfied with the performances of wholesalers as their activities ultimately aim at earning profit without any significant benefit in terms of price.

Agents

Although agents are not to be considered as one of the channel members, their significance in promoting the channel for distribution of coconuts cannot be underestimated. Agents include brokers of traders, wholesalers, industrial users and commission agents located in secondary market centre. The climbers also act as agents. Most of the big traders, including wholesalers and industrial users, engage agents to facilitate procurement of coconuts in villages. Inter-state wholesalers have agents for distribution of coconuts. As cooperative marketing bodies and regulated markets facilitate better marketing of coconuts by farmers they are referred to as agents of farmers.

Procurement Agents

They include village brokers and commission agents. They act as agents of traders and wholesalers. Through them, traders and wholesalers execute the harvesting schedule. Further, price negotiations and other terms and conditions regarding purchase and sale of coconuts are administered among farmers by these persons. They are paid depending on the quantum of transaction effected through them.

Commission Agents

Among procurement agents, commission agents occupy an important place. They are found in local and secondary market places. They collect purchase orders from buyers of coconuts and accordingly procure coconuts from sellers (mostly from farmers bringing dehusked coconuts to market centre) on cash-and-carry system. They receive their commission / brokerage from both buyers and sellers of coconuts depending on the volume of transaction.

Distribution Agents

They mostly deal in inter-state trade transactions. It is understood that there are about 60-70 such agents in Tamil Nadu. As they have trade contacts with inter-state coconut buyers, they receive purchase orders and place them with traders and wholesalers. They supply information relating to reliability, credit worthiness, etc. of the inter-state traders and agents. They also procure coconuts on behalf of inter-state traders and agents from the local traders and wholesalers.

Farmers' Agents

Pollachi Agricultural Produces Cooperative Marketing Society acts as an agent of coconut farmers. They arrange for better marketing of coconuts and copra. The role of organised marketing bodies in regulating the buyers of coconuts especially traders needs no special mention. However, their role in coconut trade is far from satisfactory. Dominance by traders, inefficient management and supervision, absence of committed officials and employees, irregularities in price fixation etc. are the major reasons for tardy performance of these organisations.

Sale Practice of the Sample Respondents

Sale practice comprises of wholesale, retail, direct sale to customer and mixed practice. The following table shows the practices of the sample respondents in the study area.

Table 5.6
Sale Practice

Category of the farmer	*Sale Practice*								*TOTAL*	
	Wholesale		*Retail*		*Direct to consumer*		*Mixed*			
	No.	*%*	*No.*	*%*	*No.*	*%*	*No.*	*%*	*No.*	*%*
Small	48	47.1	14	13.7	38	37.3	2	2.0	102	100.0
Medium	76	74.5	4	3.9	5	4.9	17	16.7	102	100.0
Big	76	74.5	2	2.0	2	2.0	22	21.6	102	100.0
TOTAL	200	65.4	20	6.5	45	14.7	41	13.4	306	100.0

χ^2 value is 89.294 which is significant at 1% level

The table 5.6 clearly depicts that a good majority (65.4%) of the respondents were following wholesale practice due to avoiding herculean task of selling in small portions, to make lump sum amount of returns, minimize the wastage, lack of manpower to do sorting and selling etc., Nevertheless 14.7 percent of the respondents taken the challenge of direct selling to consumers. It is witnessed from the survey that majority of the small farmers are undergoing direct sale to consumer. This may be due to the fact that small numbers of production, capitalizing the maximum benefit, avoid overhead costs like loading, unloading, transportation etc. The calculated chi square value confirms that there is no significant relationship between the category of farmers and their sale practice.

Table 5.7

Channel Member Choice of Coconut Farmers

S.No.	*Channel members*	*No. of Respondents*	*Percentage*
1	Coop. Marketing Society	84	27.4
2	Regulated Market	01	0.4
3	Traders	101	33.1
4	Commission Agents	43	14.0
5	Wholesalers	42	13.7
6	Local Marketers	16	5.2
7	Industrial User	19	6.2
	Total	306	100

A simple majority of the respondents (33.1%) has chosen traders as their channel member followed by cooperative marketing society (27.4 %). Flexibility in trade operations, advance payment, farm gate purchase, less cumbersome trade process might have influenced the farmers to go behind the traders. Standardized practice, warehousing facility, advance against the pledge, reasonable price, open and free sale process might have attracted the farmers to have trade relationship with the cooperative marketing society. Nevertheless, rest of the non-institutional agencies like commission agents, wholesalers, industrial user and local marketers also attracted the farmers in a notable way.

Coconut Price

A close review in this regard revealed that the price of coconuts is fixed on the basis of the price of coconut oil. In other words, the coconut oil prices tend to determine coconut prices. The coconut oil price is in turn, largely determined by a few big merchants who buy nearly 50 per cent of total production of coconut oil in the country (Markose, V.T. 1992). As the coconuts produced in Kerala State stand first in quality, they always enjoy the highest price (CDB 1985).

Since the prevailing prices of coconut and coconut oil, especially in Kerala, are published in the daily newspapers and by the Coconut Development Board (Kochi), traders find no difficulty in fixing their prices. The price, as quoted in newspapers, is taken as the ruling price. Marginal changes in price-upward and downward are often conditioned by supply and demand positions.

The prices of coconuts in Tamil Nadu generally move in line with the prices prevailing in Kerala. However, a significant portion of coconuts produced in Tamil Nadu are being sold out to other parts of the country such as Gujarat, Maharashtra and Uttar Pradesh.

The demand and supply position in these upcountry centres is also a determining factor in fixing the prices. The farmers do not seem to have a say in determining the prices for their coconuts as they are neither organised nor backed by formal organizations (Satheesh Babu, K. 1996). Further, coconut is under free trade and no organised body is effectively manipulating the demand or prices in the market.

Seasonal Variations in Coconut Price

The normal price trend of coconuts and coconut products over a period of years in Tamil Nadu indicates that the price record a downward trend from January to June, the lowest price prevailing in April-May. Thereafter, the price tends to improve till December, the highest prevailing in November (CDB 1997). Such a trend coincides to harvests of coconuts. This trend had also been observed in the study area. The income of the farmers from coconut trees tends to increase during periods of high production or during high price for coconuts. Farmers, therefore, experience seasonal variations in their income from coconut trees.

Table 5.8

Seasonal Variations in Income from Coconut Trees

Income Level	*Farmers under different categories*		
	Big	*Medium*	*Small*
Sale proceeds of coconuts			
High	March-April	March-April	March-April
Moderate	Oct-Dec	Oct-Dec	Oct-Dec
Low	June-July	June-July	June-July

Sale proceeds of by-products			
High	-	-	-
Moderate	Jan-Dec	Jan-Dec	Jan-Dec
Low	-	-	-
Price of Coconuts			
High	Oct-Dec	Oct-Dec	Oct-Dec
Moderate	June-July	June-July	June-July
Low	Apr-May	Apr-May	Apr-May

(*Source:* Subburaj. B., 2004)

The price, as obtained by farmers for coconuts, differs among regions and buyers due to variations in quality and marketing cost. The farmers' gross price for 1000 coconuts ranged from Rs.3983 to Rs.4180 while the maximum price for coconuts in Kerala was Rs.5150 per 1000 nuts (CDB 2006). Thus there is a wide variation between the prices prevailing in Kerala and in Tamil Nadu. This apart, the farmers' average gross price (for 1000 coconuts) realized through various channel members also varies. Thus, farmers' price of coconuts varies among regions and buyers.

Table 5.9

Average Market Price per Coconut for Periphery and Grove Variety during Off and Flush Season

S. No	*Particulars*	*N*	*Minimum*	*Maximum*	*Mean*	*Std. Deviation*
1	Flush season-Periphery	174	3.00	4.00	3.3678	.4836
2	Flush season-Grove	192	2.00	4.00	3.5521	.5192
3	Off season-Periphery	174	3.00	5.00	4.2989	.4836
4	Off season-Grove	192	3.00	5.00	4.4687	.5780

The survey results clearly state that the farmers fetch maximum of Rs.4.00 for both the varieties during flush season. The mean price for grove is slightly higher (3.5521) than periphery (3.3678). Indeed it is a common knowledge that the prices are generally near to the ground during flush season on the contrary the prices are sky-scraping in off season. It is clearly understood from the survey that irrespective of the variety the minimum (Rs.3.00) as well as maximum (Rs.5.00) price is same during the off season.

Satisfaction Over the Price

Better price for the products is the only satisfying and motivating factor for the producer. An attempt has been made to know the farmers satisfaction over the price offered by the respective buyers and the survey results has been furnished hereunder.

Table 5.10

Satisfaction over the price offered by the buyer

Educational Qualification	*Response*						*TOTAL*	
	Satisfied		*Not satisfied*		*Neither satisfied nor dissatisfied*		*No.*	*%*
	No.	*%*	*No.*	*%*	*No.*	*%*		
Pre-matric	44	61.1	20	27.8	8	11.1	72	100.0
Matric	62	49.2	52	41.3	12	9.5	126	100.0
UG	26	48.1	22	40.7	6	11.1	54	100.0
PG	16	72.7	6	27.3	-	-	22	100.0
Illiterate	14	43.8	8	25.0	10	31.3	32	100.0
TOTAL	162	52.9	108	35.3	36	11.8	306	100.0

Chi-Square Test

	Value	df	Sig.
Chi-Square	21.354	8	**

**: Significant at 1% level

It is clearly understood from the table 5.10 that a good majority of the respondents (52.9 %) were satisfied over the price offered by the respective buyers followed by 35.3 percent of the respondents were not satisfied with the price. It is interesting to note that, among the well educated category a good majority of the respondents (72.7%) were satisfied with the price. This may be due to the fact that these educated respondents may know the market conditions and prevailing market rate for their products. The calculated chi square value is 21.354 which is higher than the table value of 20.090 at 1% level of significance, hence it can be concluded that there is a significant relationship between these two variables.

Table 5.11
Opinion about Price

No. of years in Coconut Cultivation	*Response*				*TOTAL*	
	Institutional		*Non-institutional*		*No.*	*%*
	No.	*%*	*No.*	*%*		
Up to 10 yrs	60	63.8	34	36.2	94	100.0
11-20 yrs	86	69.4	38	30.6	124	100.0
Above 20 yrs	50	56.8	38	43.2	88	100.0
TOTAL	196	64.1	110	35.9	306	100.0

Chi-Square Test

	Value	df	Sig.
Chi-Square	3.516	2	Ns

NS: Significant at 5% level.

It is inferred from the table 5.11 that a good majority of the 196 respondents (64.1 percent) opined that the price offered by the institutional agency is better than that of the non-institutional agencies. This may be due to the fact that institutional agencies like cooperative marketing society and regulated markets are supported and governed by the government and moreover these institutions always in the side of farmers whereas the non-institutional agencies are profit centered and tries to exploit the farmer community as much as possible. Nevertheless about 110 respondents (35.9 percent) have favoured their opinion towards non-institutional agencies. Chi square value confirms that there is no significant relationship between these two variables viz., years of experience in coconut cultivation and opinion towards price offered by the institutional as well as non-institutional agencies.

Table 5.12
Reasons for Preferring Non-institutional Agencies

S. No	*Reasons*	*No. of Respondents*	*Percentage*
1	Better price	6	5.5
2	Pre and Post Services	18	16.4
3	Advance money	40	36.4
4	Farm gate sales	23	20.9
5	No cumbersome process	23	20.9
	Total	110	100.0

Amidst multiple discrepancies are associated with the non-institutional agencies, a notable size of the (35.9 %) respondents (Table 5.12) chosen non-institutional agencies for their marketing operations. Further analysis clearly expresses, a simple majority of the respondents (36.4 %) opined that advance money is the primary reason for opting non-institutional agencies followed by farm gate sales and no cumbersome process are the reasons felt equally by the 20.9 percent of the respondents. Nonetheless 5.5 percent of the respondents opined that better price was offered by the non-institutional agencies.

Need for Price Stabilization for Coconut and its Products

The price of coconut and its products, particularly fresh nuts and copra, are normally governed by the prices of coconut oil in the terminal markets. Which in turn are subject to heavy influence by the overall availability of other fat and oil in the country as well as their price behaviour in the market. As a result of continuous price fluctuations, premium price, short supplies etc., the coconut oil is slowly loosing its pre-eminence in many end-users and consequent erosion of its traditional markets. This trend is ultimately leading the growers as well as consumers into hardships. Hence, the necessity to introduce a viable price stabilization scheme with the basic objective of minimizing price fluctuations and maintaining a stable price range for coconut oil which would besides being remunerative to coconut growers, facilitates the re-entry of coconut oil into its traditional markets and promote new end-use products. These objectives could be achieved by operating a carefully managed supply management scheme, with coconut oil as the base.

MARKETING OF COCONUTS AND ITS PRODUCTS IN TAMIL NADU

Coconut

Out of the total production of 4968 million nuts (2007-08) about 47 percent is converted into milling copra and desiccated coconut, 10 percent for household and religious purposes and 10 percent as tender coconuts. Thus there is a sizeable market surplus 33 percent which is now being transported as raw coconut itself to the up-country markets in partially dehusked form. It ensures safe transport and also resists the climatic fluctuation. It also exploits the human psychological factor of appreciable sizes in appearance.

Copra

Copra is the dry form of the kernel. This process can be done either by sun drying or by kiln drying or by indirect hot air dryer or by solar dryer. So far as Tamil Nadu is concerned, sun drying is the popular method as the sun shine is ensured for about 9-10 months in a year. However, during the rainy season, copra is dried by means of smoke kiln in which direct head and smoke come in contact with the kernel, resulting in the inferior quality of copra produced and thereby deterioration of coconut oil.

There are two types of copra manufactured from coconut i.e., edible copra which again is categorized into two i.e., ball copra and cup copra. The other variety is milling copra. However the practice of making edible copra is not a vogue in the state despite the fact that the copra available in the region is sweet and good for edible purpose. Possibilities for conversion of a portion of the coconut into edible copra could be explored.

As it stands, milling copra production is the major activity concentrated in various pockets of Tamil Nadu. Therefore, more attention is to be given for improving its quality.

Coconut Oil

Coconut oil is produced by crushing the copra. In recent years, there are number of units coming forward to install more expellers. It is reported that there are about 100 medium and big oil milling units in Kangeyam, Erode District.

It is incidental to mention here that a recent study conducted in Madhya Pradesh revealed the fact that the main procurement centre of coconut oil is partially shifted from the traditional market of Cochin to Kangeyam. The coconut oil price in the country is at present determined by the price quoted at Cochin market. However, the price at Kangeyam always rules below the Cochin market, resulting in an added advantage to the traders of the up-country markets. The present system of marketing of coconut oil is that oil in bulk containers are sent to the up-country markets and from there re-packed into small packs.

Desiccated Coconut

There are about 12 desiccated coconut manufacturing units in Tamil Nadu with an annual production of around 6,000 metric tones which is about 30 percent of the total production of the country estimated at 20,000 metric tones per annum. There is a tremendous scope for boosting the demand for the product both in the domestic market and the Gulf and Far East.

Tender Coconut

Out of the total production of 4,346 metric tones, 10 percent is consumed as tender coconut. There is no in-built system at present prevailing for the marketing of tender coconut. However, the brokers directly harvest nuts from the gardens and sell by themselves in many areas. There are wholesale merchants also who collect the nuts from the gardens and transport it in truck load to the major towns and cities from where again it is marketed through wholesalers and retailers. The major items which could be made out of coconut are the following:

Coconut Cream, which is a concentrated form of milk extracted from the wet matured kernels, is at present available in the market. Similarly, coconut milk spray dried coconut powder, coconut honey, Nata-de-coco etc., could also be produced. From the shell, Shell charcoal, activated carbon, shell powder etc., could also be produced.

MARKETING STRATEGIES FOR COCONUT AND ITS PRODUCTS

The scenario of market, both in domestic and export trade, perceptibly indicates the market share of coconut oil is declining, as there is a stiff competition, especially from palm oil and soybeans oil, on prices. Diversification efforts have met with limited success, and this trend in coconut economy will continue. How to ameliorate the situation to have sustained growth in coconut is a challenge. In this scenario, strategic market research would be essential to link the consumer, customer & public to the marketing through information, coupled with generation refinement and evaluation of marketing actions, monitoring market performance; and understanding of marketing as process. Market research could be helpful in market planning; where aim is to devise a plan concerning product, advertising, and related matters, to help assure marketing success; problem solving – for addressing specific issues and monitoring – to measure what is going on to ensure that all is well and to spot possible problems. Scenario in coconut is that share of coconut oil is reducing. Reason could be higher price, poor packing, less awareness about the product. The well-thought marketing strategies could be to promote the product more vigorously, providing information on health aspect to the targeted customers. Who could be the target customer, may also vary depending upon the socio-economy of population. Thus, market research would help in decision making process.

Coconut industry is in the process of bringing new products in the market. New co .onut products, as in the case of any other technology driven new product, may become a failure in the market due to improper market access. The failure of such new products is often attributed to the lack of understanding of the customers need, requirements, desires, and priorities. Properly designed products with advance market data will reduce the risk. New product ideas can be had by identifying proper focus groups, interviews and questionnaires, generating a product concept from the customers' perspective etc. The critical issue that influences the ultimate success of a product of failure must be addressed. In this process the market forecasts can be developed. Therefore, to tap on emerging market, to access the market for various products, to identify the need of consumer for refining the product-market research have to be given priority which should analyze: What are the advantages and benefits of the new product as compared with existing products? What product attributes is most important in the buying decision? What makes the product unique compared with their current product? How satisfied are consumers with their current product? What are the characteristics of consumers who are most likely to purchase the new product? Does the product meet the requirements and desires of the target market? Will the new product cut into sales of the existing products? These and similar other issues warrants for an in depth marketing research.

Marketing operation becomes more effectively when Competitive Marketing Intelligence and Marketing Information System work in synergy. Competitive Marketing Information comprises collection and storage of data, analysis and interpretation of data and the dissemination of intelligence. This process provides information about competitors, product quality, competitors' attitude towards new product, competitors' attitude towards bottom line etc. It also provides information about the strategic nature of the competitors market, information about market share, market trend as well as on competitive distribution channels. The competitive marketing Information system measures competitive flexibility in marketing strategy. It assembles all pertinent marketing data about a product line from both external and internal sources and assembles the data so that they can be integrated into any combination to help solve a problem or to monitor ongoing marketing activities. In other words Market Information System is a data base tool to help in marketing decision making.

There has not been a well designed marketing strategy in India for the coconut products. This is more so in the case of Indian coconut products as well. The Coconut Development Board is on the path of bringing a major shift in the marketing strategy in tune with the technology advancement in product development as well. New technology driven products, orchestrated with active product promotional campaigns could capture the markets for selected products. The entrepreneurs with their products are brought under the same roof in national and international exhibitions and fairs in a strategic manner and given chance to expose them to public, which has helped to get better access. In this process, entrepreneurs develop networking and market tie ups with prospective companies.

The demand for health foods and natural beverages has been increasing at an alarming rate world-wide. Coconut industry has the potential to benefit from the changing food habits through product diversification. Coconut has to be projected not as an oil yielding seed but as a dietary fruit endowed with considerable health benefits. The functional drinks from coconut have potential for growth in the world market, as the current markets are attracting health conscious groups. Coconut products have high medicinal properties. Studies reveal that the fatty acids present in coconut products provide energy and raw material for anti-microbial fatty acids and mono-glycerides. Studies undertaken in and outside the country on the health aspects of coconut oil certified it as beneficial dietary oil and which does not contribute to any of the risk factors to the human body. The Lauric acid, component of coconut oil is considered as an ideal dietary fat due to its anti-microbial benefits. Human body synthesizes monolaurin from lauric acid which possesses antiviral, antibacterial and anti-protozoal properties. Studies have indicated its effectiveness in lowering the vital load of HIV, which means less AIDS presence. Tender coconut water, a delightful natural drink, rich in minerals and vitamins, has many medicinal properties. Since the health conscious population is looking for natural health drinks, there is ample scope for popularizing this product as health drinks, sport drink etc.

In the marketing strategy, market modification and product

modification are important aspects. Market modification involves converting non-users to users, entering in new market segments, winning competitive customers etc. In product modification, quality improvement and product feature improvement are relevant factors. The buyers prefer attractive well presented packaging at easily accessible locations. This is applicable to edible as well as non edible products. This area requires tangible improvement in producing countries for competing in the international markets. Elite class people with high purchasing capacity are least bothered on the cost factor; instead, they are more conscious on the attractive packaging which can add fillip to keeping the quality as well. At the same, higher production cost of the product coupled with the high cost involved in attractive packaging is a major discouraging factor to attempt to costly packaging by the entrepreneurs. Better market movement of the product can solve this problem. Coconut products are gaining increasing consumer demand in areas where coconut is not grown. Tender coconut, desiccated coconut, coconut chips, coconut water based products are gaining momentum in the non-producing states. In the proposed marketing system adequate attention should be given to introduce such non-traditional products to the potential consumers in the non-producing areas.

Problems in Coconut Marketing

As the socio-economic environment to the society, suitable changes in infrastructure will have to be made. Slowly but surely, the villages have now become a priority item in the national agenda. The green revolution has made the village markets boom. The marketing of agricultural and non-agricultural products assumed new dimensions. Yet the marketing aspect of rural economy has remained neglected.

The consequence has been that producers are being subjected to exploitation. The cooperatives too have not been able to deliver the goods. The traditional marketing system has become obsolete. This has created a gap which needs to be filled. The middleman rule supreme in village markets. An overwhelming number of small farmers cannot individually afford to make satisfactory sale arrangements for their produce. They have to face many difficulties in the process of marketing not only are dubious practices adopted by the middleman. They make the sellers pay variety of extra charges. A favorable or conducive atmosphere through well structured system has to be created to protect the farmers' interest.

Marketing is a part and parcel of modern productive process, the part of the end that gives point and purpose to all that has gone before. Marketing thus begins at the farmers' field. In the case of coconut it includes all such activities as (1) collection of surplus from the individual farmers (2) Transportation (3) Processing coconut into copra and copra into oil cake (4) grading and standardisation (5) Warehousing (6) Packing (7) Bringing the buyers and sellers together and (8) Sale to the ultimate consumer. Both processing and marketing of coconut are intimately connected with each other, sometimes processing serves as an adjacent to the marketing operations and vice-versa. It is because of this close relationship between the two activities that the cooperatives dealing with coconut, undertake both processing and marketing simultaneously.

Problems are part and parcel in agricultural marketing. Coconut marketing too is facing numerous problems. The problems as perceived by the respondents are discussed hereunder:

Table 5.13

Problems in Coconut Marketing

S. No	*Problems*	*No. of Respondents*	*Percentage*
1	Unethical marketing practices	10	3.3
2	Absence of stabilized price	80	26.1
3	Increased cost of marketing	82	26.8
4	Absence of standardization and grading	34	11.1
5	Less facilities for coconut processing	78	25.5
6	Ineffective institutional agencies	14	4.6
7	Forced sale	8	2.6
	Total	306	100.0

It is evident from the table 5.13 that 26.8 percent of the respondent opined unethical marketing practices is one of the major problems faced by them followed by absence of stabilized price (26.1%) and ineffective institutional agencies (25.5%). This may be due to the fact that the study area is still dominated by the non-institutional agencies and their prime motto is to capitalize the market opportunity not the welfare of the coconut farmers. It is very distrusting factor that institutional agencies like cooperative marketing society and regulated market are also ineffective in the study area.

Apart from the above, the below mentioned problems were identified through informal discussions and focused group discussions held among the coconut growers, market functionaries and institutional officials in the study area.

Price Instability

The instability in coconut oil prices and consequent instability in copra and coconut prices pose serious problems in coconut marketing sector. The wide fluctuations in the average annual wholesale price and average monthly wholesale prices could be attributed to more than one reason. While the prospects of improved supplies of oil seeds and oil following favorable weather favour the lowering of prices, the apprehension of unfavourable weather pushes the prices forward.

Petite Number of Marketed Surplus

The marketed surplus of nuts per holding is very low; it is not worthwhile to carry over long distance for disposal. Again with such a small lot, the farmer finds himself at a disadvantageous position to bargain for fair price with merchant in assembling market. Other compelling factors like pre-harvest contract and inaccessibility to market give rooms for sales at gardens. Under these situations, the middlemen abuse their positions by doing various types of malpractices. They collect samples and make some sort of arbitrary discounts on agreed rate under the pretext of undersized and immature nuts.

Stock and Prioritization

Among others forced sale is felt as the first and foremost problem followed by non-availability of organised credit and ineffective organised marketing institutions. Moreover, it is revealed that all the above problems are interdependent and, cumulatively they affect the farmers' income from coconut trees. The inventory of marketing problems is discussed here under.

Strained Sale

Dearth of staying capacity among farmers largely results in forced sales, which by and large, contributes to prior debt. The coconut farmers mostly borrow from coconut traders to meet production and consumption expenses. The other equality important reasons are absence of storage facilities and fluctuations in price. These factors force the farmers to sell their coconuts on the farm immediately after harvest. The problem of forced sale is more common among lessor farmers as two-thirds of them lease their coconut trees for want of organised credit to meet their production and investment requirements. If the problem of credit is not given due attention, there is a possibility of opportune farmers also becoming lessor farmers.

Absence of Financial Support from the Organized Financing Institutions

Non-availability of organized credit is a severe problem resulting in indebtedness and diminishing bargaining power among coconut farmers. The need for short term production credit, especially during off seasons and investment credit during flush seasons (summer days) has forced coconut farmers to raise temporary loans and advances from private sources. The need for credit is very acute during the fourth and fifth year of the coconut crop, when there would be no income, especially to those who raise coconut trees in grove as mono crop, because they can neither raise inter-crop nor derive yield from the coconut trees even to meet their normal production and consumption expenses. However, a few opportune farmers have utilized credit from organized financing institutions. Commercial banks do not provide short term and long term credit to coconut farmers against their standing coconut trees. They extend crop loan mostly on personal surety and security of anticipated crop, and pledge of collaterals and valuables. Long term credit is provided against mortgage of land and other immovable properties.

Ineffective Organised Marketing Institutions

The organised marketing institutions include cooperative marketing societies, and regulated markets which are available in the study area. Amidst existence of organized marketing institutions, they do not make any arrangements for procuring coconuts and copra from farmers through out the year. Although coconut and copra have been declared as notified crops, they are not properly rooted through organized marketing institutions. The reasons for the failure of the organised institutions to solve the marketing problems of the coconut farmers are : dearth of facilities, absence of buyers of copra due to increased cost of transportation, less advantage in price realization, etc.

Existence of Traders' Coterie

Coconut traders have both formal and informal associations at all levels led by big and wholesale traders. The big and wholesale traders play a critical role in the distribution channels for coconuts. They never fail to protect the interests of their traders. There are instances where the trading community, through their solidarity, have shattered several informal associations of coconut farmers, and led the coconut farmers into severe hardships.

Absence of Stabilized Price

The price of coconut is high during off seasons and low during flush seasons. However, such a trend does not prevail as a rule. Variations in size, quality and location differences of the farm and the plight of the farmer are the limiting factors in fixing a base price for coconut. The traders, by taking advantage of the above limiting factors, put the farmers in a state of desperation, exploit them and boost their own profit.

Lack of Market Information

Adequate market information, particularly the price of coconut, is not available. Although the price of coconut is determined on the basis of copra price, farmers find it difficult to understand the methodology involved in fixing the price based on the oil recovery percentage of their coconuts. Further, as the bargaining power of farmers is very weak, they have no hold over their coconuts. Even the farmers of Pollachi region, with relatively high quality coconuts, are unable to fix the price of their coconuts. The market information published by the Government bodies and newspapers has benefited largely the traders/processors and not the farmers.

Be Short of Facilities for Processing of Coconuts

Indigenous oil crushing units of small scale industries exist in all the places of Anamalai block and Pollachi but these processing units prefer large scale traders and wholesalers due to advantages in their scale of operations. Further, copra making, either as a cottage or as a small scale industrial activity is not found among traders and farmers, due to absence of local market for copra, increased cost of transportation, differences in quality of coconuts and cost of credit bill collections from buyers of copra.

Desiccated coconut powder units also exist. But they too face problems like non-availability of required quality of coconut, high cost of procurement, fluctuations in price of coconuts, paucity of funds, difficulty in maintaining output quality, and competition from units located in nearby States like Karnataka. We can see a large number of small scale coconut fibre units. But they prefer to procure their raw material (coconut husks) from traders as it is available in bulk from them. Traders or coconut fibre units do not buy coconut husk from the farmers due to high cost of assembling.

Shoot up of Cost of Marketing

Cost of marketing increases due to increase in wages of tree climbers, cost of assembling of coconuts in the farm during harvests, the cost of tender nuts consumed, cost of loading and transportation and cost of peeling coconuts. The labourers who are employed in coconut harvest are to be provided nuts of superior quality at free of cost. While selling coconuts either through traders or commission agents within and outside the village, quality and quantity allowances in terms of coconuts are to be provided. Another interesting fact is that the cost of marketing increases due to damage of coconuts. The proportion of spoilage is higher to a farmer who wishes to sell his coconuts directly in market places than to those who wish to sell coconuts through traders on the farm itself. The increased spoilage is primarily due to deliberate harvest of immature nuts by tree climbers who, in fact, act as agents and brokers of big traders in villages. The farmers cannot antagonize tree climbers as they are in high demand and possess special skills. Farmers, in fact, need to maintain a good rapport with them by ignoring their faults and frauds.

Absence of Standardisation/Grading

Absence of standardisation and grading of coconuts is a serious marketing problem, which results in poor price for coconuts. The traders, while procuring husked coconuts at farmers' fields, do grade them based on size and maturity. Maturity of coconut is judged by its weight and water content. The inferior nuts are separated. This is done mostly be tree climbers who have expertise acquired through experience. Grading on the basis of size of the nut is not done when the coconuts with husks are sold. With regard to dehusked coconuts, four grades of coconuts based on their size are common in villages. The number of dehusked coconuts that could be packed in a single gunny bag of standard size (wheat or sugar gunny) serves as a yardstick. The grades based on size of the coconuts are: a gunny bag full of 150 coconuts;

a gunny bag full of 200 coconuts; a gunny bag full of 250 coconuts; and a gunny bag full of 300 coconuts; However, price for each grade of coconuts is not uniform throughout Tamil Nadu at any given time due to variations in outturn of coconuts, system of sale adopted by farmers, etc.

Recommendations

The following recommendations may be implemented with the help of concerned Government authorities, non-governmental organizations and farmer's organizations to tackle the coconut growers marketing problems.

Formation of exclusive coconut producers cooperative marketing society

The statistical facts confirm that Coimbatore district stands first in area, production and productivity of coconut. The existing Pollachi Agricultural Producers Cooperative Marketing society is a general society which deals with number of agricultural products. Coconut is one among them irrespective of its bulkiness moreover it is undertaking coconut marketing only because of market intervention scheme of National Cooperative Marketing Federation (NAFED). The scheme is not a permanent affair; it may extend for three or four months with the aim to stabilize the market and price. Indeed it is common knowledge that with in the short span of time, it is hard to set right all things. Hence, formation of exclusive coconut producers cooperative marketing society in the lines of turmeric producers' cooperative marketing, groundnut producers cooperative marketing society will definitely solve the problems of coconut growers in the study area.

Provision of Organised Finance through Government and Institutional Agencies

Among the remedial measures the most favoured one is provision of organised finance to farmers through Government and Institutional Agencies like Cooperative Marketing Society and Regulated Markets. They require short term and long term credit, to be dispensed by organised financing institutions. The following modalities are suggested for distribution of credit.

Provision of Short Term Credit

The need for short term credit may be fulfilled on the basis of annual average sale proceeds of coconuts which urged to meet production and consumption expenses. Farmers feel that provision of credit against coconut trees is more secure than a crop loan provided against the anticipated crop.

Provision of Long Term Credit

The need for long term credit is keenly felt by farmers for the following reasons: (i) to get rid of the problem of leasing and or re-leasing coconut trees; (ii) to bring improvements in the farm investments especially irrigation sources; (iii) to facilitate occupational diversification, including taking up cottage and small scale coconut-based processing / industrial activities; and, (iv) to redeem prior debts.

Price Guarantee Scheme

Farmers suggest price guarantee schemes as an effective measure to ensure reasonable price for coconuts. Such a scheme will regulate seasonal fluctuations in price of coconuts coconut products, protect farmers from traders ring and promote copra processing on the model of cottage and village industries. The farmers also suggest a base price for coconut, to be fixed on the basis of oil recovery percentage of coconuts like price fixed on the basis of sugar content in the sugar cane.

Establishing Integrated Coconut Industries

Besides promoting coconut-based cottage and village industrial and small scale units among farmers in villages, the need for large scale integrated coconut processing complexes is felt, as it will facilitate disposal of coconuts and copra without depending on any channel members, particularly the traders. While discussing the constitution and management of such integrated coconut processing industries, farmers favour the idea of state patronage.

Revamping of Organised Marketing Institutions

Farmers appreciate the services of the regulated market but point out its deficiencies too. The farmers, by and large, do not recognise the existence of cooperative marketing societies. However, realizing the potential of cooperatives and the regulated markets in protecting farmers from traders, they suggest revitalization of organised marketing bodies.

Farmer's Associations

Farmers do understand the importance of corporate entity

under the provisions of law. But such an entity, the farmers feel, often endangers their autonomy and independence. The case on hand is that of cooperatives. Hence, they wish to have self-regulated self-help (informal) associations with patronage from the Government and banking institutions.

Sum Up

The coconut industry is growing in terms of production. However, its share in oil and fats trade has consistently declined in the past four decades. Vast growth opportunities remain for the coconut industry, but the marketing strategy needs a reorientation to suit to emerging trend. Awakening and alertness on diversification of coconut, with a motive to recapture the market have to be provided priority. The best option with vast growth opportunities for product diversification and value addition exists. Despite this vast potential, the industry can flourish only through strategic initiatives and synergy among the organised and unorganised market outlets function. Strategic marketing has to include product diversification, marketing intelligence backed by market research to widen the market base in different coconut growing region. Cooperation from the organisations and upfront approach is needed in partnership mode to address the issues in penetrating the existing markets.

Since the existing marketing networks are dominated and controlled by private marketing functionaries, farmers are deprived of the major chunk of the consumers' price for coconuts. Farmers know each and every disadvantage of sale through the existing channel members. However, they depend on them by compulsion. Non-availability of organised credit for production, marketing and consumption purposes, absence of organised marketing institutions and difficulties in the adoption of off-farm value addition, processing and marketing technologies on coconuts are the major reasons for the 'forced sale' of coconut either through lease practices or on-farm sale of coconuts to marketing intermediaries. They, therefore, demand rejuvenation of organised marketing institutions such as cooperatives and regulated markets and or promotion of a network of micro-enterprises of coconut farmers facilitating for collective bargaining and direct marketing of coconuts in places where the organised marketing institutions are defunct.

6

Coconut Growers' Perception about the Present Marketing Functionaries

Introduction

Marketing channels for coconut and coconut products are comparable to those of other agriculture product. The channels can be broadly classified in to two namely Institutional Agencies and Non-Institutional Agencies. Cooperative Marketing Societies, Government run Regulated Markets and Market Committees are coming under the category of institutional agencies. Rest of all private market players is occupying the state of non-institutional agencies. The private marketing channels linking producers and consumers consist of intermediaries namely traders, commission agents, wholesalers, local marketers and retailers with in the state. In the case of inter-state trade, it passes from wholesale dealers/ commission agents of the state to the wholesale dealers/ commission agents in other state.

The marketing of coconut in Tamil Nadu is routed through the cooperative marketing societies, regulated markets, and wholesaler holders to local markets, oil miller and other means. Some of the coconut producers sell their produce directly to the oil miller if they have a strong financial and business relationship with the oil mill entrepreneurs. The lessee of coconut trees sends his produce to the retailers either through wholesaler holder or local market or commission agents. Indeed it is common knowledge that there is no much difference in the net price received by producer and lessee. Hence the produce is moved through wholesaler to local market and retailer to reach the ultimate customer. The wholesaler plays an important role in promoting coconut markets in Tamil Nadu. They generally procure coconuts from the producers in large quantities and also sell them in large quantities. It can be seen that wholesalers usually sell coconuts to oil millers, retailers and traders in other states. The wholesalers do not continue themselves to one particular market due to capitalize the market opportunities. The retailers purchase coconuts from local market/commission mundy. They sell their coconuts to small shops, household consumers, hotels, and the like.

In the study area, both institutional agencies and non-institutional agencies are participating in coconut marketing operations. Preference Score Analysis was employed with appropriate statistical tools like chi-square test, ANOVA and t. test to know the preference of the sample respondents towards the functions and role of institutional and non-institutional agencies in production and marketing of coconut.

Prevailing Market Outlets for Coconut in the Sample Area

The prevailing market outlets for coconut in the sample area can be basically grouped in to two major categories viz., institutional agencies and non-institutional agencies. Under the first category there are two institutions namely Pollachi Agricultural Producers Cooperative Marketing Society Limited and Anamalai Regulated Market. Under the second category there are commission agents, traders, wholesalers, local marketers and industrial users. The functions and role of these agencies in coconut marketing is discussed hereunder:

Institutional Agencies

i) Pollachi Agricultural Producers Cooperative Marketing Society Ltd

To a farmer who is conscious of his economic interest, marketing co-operative is as important as a credit co-operative. Theoretically assumed advantages like greater holding power, high price, reduced operating cost, orderly marketing, improved bargaining, elimination of trade abuses, standardization and grading, better methods of selling, elimination of middlemen profit etc., are the main cause of the existence of cooperative marketing societies. To provide these benefits to the farmers, the Pollachi Co-operative Marketing Society limited was registered and functioning under the Tamil Nadu Co-operatives Act.

To protect the interest of Pollachi region farmers and enable them to obtain good remunerative prices for their produces, the Pollachi Agricultural Producers Cooperative Marketing Society was registered on 25.04.1933 with limited liability. It started its basic functions from the date of 19.05.1933.

The area of operation of this Pollachi cooperative marketing society confines to Pollachi Taluk, which includes Pollachi (North), Pollachi (South), Anamalai and Kinathukadu unions.

At present there are 5757 'A' class members enjoying the services of this CMS. Out of 5757, individual farmers occupy a premier position of 5664 (98%), followed by cooperatives functioning in the Pollachi area (92) and State Government.

To meet the financial needs, the Pollachi APCMS mobilizes funds in the form of share capital, entrance fee, fixed deposits, loans from central bank, State Government. The Authorized share capital of the society is Rs.50 lakhs.

Presently the functions of Pollachi APCMS are looked after by a special officer appointed by the Government of Tamil Nadu. The paid administrative staffs maintain the day-to-day affair of the society.

Functions of Pollachi CMS

The Pollachi APCMS is a general purpose marketing society, which undertakes all types of business operations for the economic and social welfare of the farmers. It undertakes marketing of member's produces like Cotton, Groundnut, Maize, Vegetables and Copra.

Apart from the basic marketing operations the Pollachi APCMS rendering its service both for the members and non-members by way of providing warehouse (2), yards (2), supply of agricultural machineries, supply of fertilizers, inputs etc.

Copra Procurement—Market Intervention Operation

Two types of copra namely milling and edible are made in India. Milling copra is used to extract oil while edible copra is consumed as a dry fruit. Edible copra is made in the form of balls as well as cups. Copra contains the highest percentage of oil (65%) compared to other oil seeds.

Since 1991-92, Tamil Nadu state has emerged as major producer of milling copra by increasing the share to over 41% from 5.6% of the total output in country. The price of coconut and coconut products has not stable in recent years due to market imperfection and prevalence of more private market operators. To safeguard the small and marginal coconut farmers, the Central Government launched a support price scheme for copra procurement. The support price will be effective only if it is backed by market interventions. The Government of India with the help of National Cooperative Marketing Federation (NAFED) has formulated a market intervention operation. Under this scheme major coconut cultivating areas are identified and procurement function is undertaken with the help of Tamil Nadu Cooperative marketing Federation (TANFED) and respective agricultural cooperative marketing societies.

The statistical fact confirms that the Coimbatore District is one of the major coconuts producing area in India and Tamil Nadu in particular. Within the Coimbatore district Pollachi region has good concentration of coconut area and production but there is no specialized cooperative marketing society for coconut and coconut products like copra, coconut oil, etc. Since there is no exclusive cooperative coconut marketing society in the region, four agricultural producers' cooperative marketing societies were selected by the NAFED for its market intervention operation for copra procurement. They are:

i) Coimbatore Agricultural Produceıs Cooperative Marketing Society
ii) Pollachi Agricultural Producers Cooperative Marketing Society
iii) Palladam Agricultural Producers Cooperative Marketing Society and
iv) Udumalpet Agricultural Producers Cooperative Marketing Society

Copra Procurement by Pollachi APCMS

Since, the Pollachi town is more approachable and well connected with infrastructural facilities the TANFED has selected the Pollachi Agricultural Producers Cooperative Marketing Society for its copra market intervention operation. Along with other marketing activities it successfully undertakes copra procurement from its members and non-members in Pollachi region with the help of NAFED AND TANFED.

The following flowchart reveals the system of copra procurement market intervention operation.

Chart: 6.1

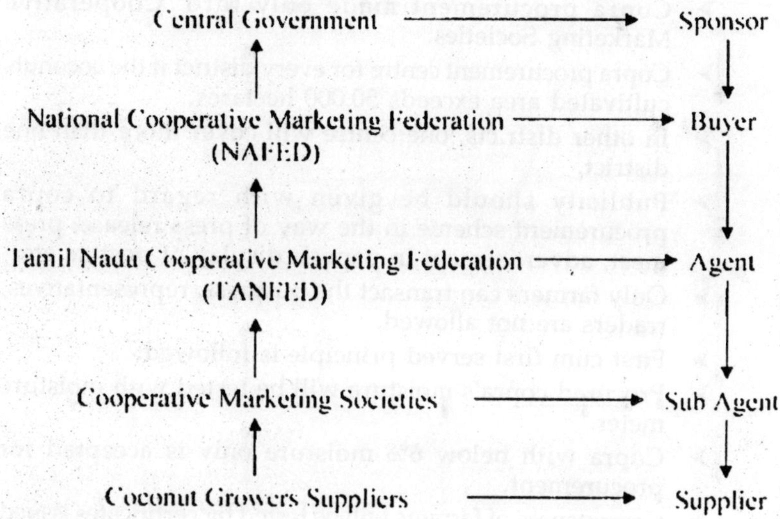

Copra Procurement Schedule of the APCMS in Coimbatore District

Copra procurement schedule by the Cooperative Marketing Societies in Coimbatore district is given below:

Table 6.1

Cooperative Marketing Society Procurement Schedule

S. No.	*Name of the CMS*	*Procurement Day*
1	Coimbatore	Wednesday
2	Palladam	Monday, Tuesday
3	Udumalpet	Thursday, Friday
4	Pollachi	
	a) Pollachi Centre	Monday, Tuesday, Wednesday
	b) Anamalai Centre	Thursday, Friday

Procedures in Copra Procurement Scheme

In order to vouchsafe and make sure that benefits of this market intervention operation should only goes to genuine coconut producers, certain guidelines have been formulated and followed by the concerned authorities.

- Copra procurement made only thro' Cooperative Marketing Societies.
- Copra procurement centre for every district if the coconuts cultivated area exceeds 50,000 hectares.
- In other districts, one centre will cover more than one district.
- Publicity should be given with regard to copra procurement scheme in the way of press release, press meet, advertisement in papers, display of posters etc.
- Only farmers can transact the business, representatives, traders are not allowed.
- First cum first served principle is followed.
- Procured copra's moisture will be tested with moisture meter.
- Copra with below 6% moisture only is accepted for procurement.
- Genuineness of farmers will be tested by certificates issued by Assistant Director for Agriculture, Village Administrative Officer and Extension Officer for Agriculture.
- Registration by the farmers only after the advertisement with regard to procurement.
- Registration only based on Identity card, which include the following particulars.
  a) Farmer's Photo
  b) Name of the Farmers
  c) Member No
  d) Permanent Address for the Farmers
  e) Survey No. of Land
  f) Coconut Cultivated Acreage
  g) No. of Trees
  h) Details about the Variety
- The CMS should maintain a separate register especially for copra procurement.
- The minimum quantity to be procured by each CMS is 1,000 tonnes per cycle.
- Only one fourth of the annual yield should be procured in anyone procurement from the registered coconut grower.

- There should be a minimum of 3 months gap between the two procurements.
- Coconut yield is calculated by the Agricultural Department Officers.
- After the procurement process, payment will be made only through A/c payee cheque.
- The CMS should daily send the statement to the Regional Joint Registrar for cooperatives showing details about copra procurement.
- 0.5% commission to the respective CMS and 0.5% of TANFED will be made by the NAFED.

Results of Focus Group Discussion

Focus Group Discussion was held with the aim to gauge the benefits and constrains in the market intervention operation and role of Pollachi agricultural producers' cooperative marketing society among the group of beneficiaries cum members in Marchinaickenpalayam village. The participants of FGD offered some recommendation to overcome the existing problems in the market intervention operation. The outcome of the focus group discussion is discussed here under:

- ✓ In search of credit facility they are forced to sell the nuts at the farm gate itself.
- ✓ For want of facilities, they are not able to convert coconut into copra.
- ✓ Indirectly traders are the real beneficiaries under this scheme in spite of all precautious.
- ✓ The coconut growers are selling their nuts before plucking due to financial constraints.
- ✓ Lengthy (6-month) business transaction is not endurable to the small and marginal farmers.
- ✓ Since the procurement operation is carried out in the marketing society premises, the overhead cost for transportation, loading, unloading and packing is more.
- ✓ The support price policy indirectly stabilizes the open market prices.

Strategies for the Development

The market intervention operation and price support scheme is meant for helping the small and marginal farmers but the scheme doesn't provide fullest benefits due to market loopholes. Price support alone will not help the small and marginal farmers but it should empower with a comprehensive support system. The Pollachi PACMS may adopt the following strategies to improve their services and to keep the members intact.

- Infrastructure facilities to be developed like drying yard, farmers rest room, transportation and proper storage facility.
- Processing activities to be initiated, particularly in the field of separating copra from coconut, oil extraction, coir industry etc.
- Public relations and member relations to be enhanced at a satisfactory level.
- Overhead cost to be minimized as much as possible with regard to transport cost, loading and unloading and packing charges by way of undertaking the procurement process at farm gates.

Preference Score Analysis

To recognize the fondness of the sample respondents towards the market functionaries preference score analysis has been made. Classification of the farmers was taken as variables for further understanding. The result of preference score analysis in this regard is discussed hereunder: The mean preference score of respondents towards Cooperative Marketing Society

Table 6.2

Mean Preference Score of Respondents Towards CMS

Classification of the farmer	*Mean*	*S.D.*	*No.*
Small	4.38	1.76	102
Medium	3.93	3.09	102
Big	4.52	2.60	102
TOTAL	4.28	2.55	306

ANOVA

	Sum of Squares	*Df*	*Mean Square*	*F*	*Sig.*
Between Groups	19.320	2	9.660	1.490	Ns
Within Groups	1964.069	303	6.482		
Total	1983.389	305			

NS: Significant at 5% level.

It is observed from the table 6.2 that the mean score of big farmers (4.52) is higher than that of small and medium farmers towards Cooperative Marketing Society. This may due to the fact that all the big farmers have a strong influence in the operation of marketing society and their capacity to transport the material to the market yard. The Analysis of variance to test the group differences is shown in the table. The F-test is significant at 5 % level which confirms that there is no significant difference among the group towards to cooperative marketing society.

ii) Anamalai Regulated Market

Regulated markets are organized by the District Market Committee under the Tamil Nadu Agricultural Produce Market Act 1959. The regulated markets are expected to counteract the malpractices prevailing in the assembling market against the interest of peasants and to provide advantages like reducing forced sales, fair price, uniform unit of provision of amenities, settlement of disputes in a fair manner etc. The Anamalai Regulated Market is the only regulated market functioning in the study area which is controlled and supervised by the Coimbatore District Market Committee. Unfortunately the Anamalai Regulated Market does not deal with coconut and it's by products. Although, the regulated market is not dealing with coconut some of the sample respondents preferred this source to market their coconut due to the nearness to the production place. The mean preference score of the respondents towards the Regulated Market is discussed hereunder:

Table 6.3

Mean Preference Score of Respondents towards Regulated Market

Classification of the farmer	*Mean*	*S.D*	*No.*
Small	.02	.14.	102
Medium	.05	.26	102
Big	.02	.14	102
TOTAL	.03	.19	306

ANOVA

	Sum of Squares	*Df*	*Mean Square*	*F*	*Sig.*
Between Groups	0.0588	2	0.0294	.835	Ns
Within Groups	10.676	303	0.03524		
Total	10.735	305			

NS : Significant at 5% level

The mean score of the medium farmers is slightly higher than that of the small and big farmers. The F- test value is at 5 % level of significance which shows that there is no significant difference among the group in the average preference score towards the regulated market.

Non-Institutional Agencies

Non-institutional agency refers the private market functionaries engaged in coconut marketing. There are five market functionaries under this classification namely, traders, commission agents, wholesalers, local marketers, and industrial user. Among the non-institutional agencies it is observed that traders have been playing a major role in all respects. The preference score of respondents towards non-institutional agencies are discussed in the following paragraphs:

i) Traders

Traders can be categorized in to three based on the activities and functions namely itinerant traders, independent traders and debt bonder traders. Each category of trader has their own uniqueness and strategy therefore, their contribution to the channel system is significant. The mean preference score of the sample respondents towards the traders is discussed hereunder:

Table 6.4

Mean preference score of respondents towards Traders

Classification of the respondents	*Mean*	*S.D*	*No.*
Small	4.86	2.00	102
Medium	6.03	5.33	102
Big	8.95	7.60	102
TOTAL	6.61	5.73	306

ANOVA

	Sum of Squares	*Df*	*Mean Square*	*F*	*Sig.*
Between Groups	904.752	2	452.376	15.050	**
Within Groups	9107.745	303	30.059		
Total	10012.497	305			

** : Significant at 1% level

With reference to the table 6.4, the mean score for big farmer (8.95) is quite higher than that of medium (6.03) and small farmer

(4.86). Thus one can conclude that there is a strong trade relationship between the big farmers and traders. Flexibility in the trade operations, farm gate sales and transport facility might have influenced the big farmers to have trade relationship with traders in spite of their membership in the cooperative marketing society. The Analysis of Variance to test the group difference is shown in the table. The calculated F -test value is (15.050) confirms that there is significant difference among the groups in the average preference towards the traders. Hence HO_6 : "There is no significant difference among the type of farmer groups in the average preference score towards traders" is rejected.

ii) Commission Agents

Although agents are not to be considered as one of the channel members, their significance in promoting the channel for distribution of coconuts cannot be underestimated. Agents include brokers of traders, wholesalers, industrial users and commission agents located in secondary market centres. The climbers also act as agents. Most of the big traders, including wholesalers and industrial users, engage agents to facilitate procurement of coconuts in villages. Inter-state wholesalers have agents for distribution of coconuts. As cooperative marketing bodies and regulated markets facilitate better marketing of coconuts by farmers they are referred to as agents of farmers. The mean preference score of the sample respondents towards commission agents are discussed below:

Table 6.5

Mean preference score of respondents towards Commission Agents

Classification of the farmer	*Mean*	*S.D*	*No.*
Small	4.69	1.66	102
Medium	6.90	5.93	102
Big	9.35	7.26	102
TOTAL	6.98	5.80	306

ANOVA

	Sum of Squares	*Df*	*Mean Square*	*F*	*Sig.*
Between Groups	1111.608	2	555.804	18.409	**
Within Groups	9148.275	303	30.192		
Total	10259.882	305			

** : Significant at 1% level.

The mean score of big farmer (9.35) is marginally higher than medium farmer and small farmer (Table 5.20). The F- test value is 18.409 which is greater than the table value of 4.676 at 1 % level of significance. Since the calculated value is greater than the table value it is inferred that there is significant difference among the groups in the average preference towards to commission agents.

iii) Wholesalers

In the channels of distribution of coconuts, wholesalers occupy an important place. They are found in regional and mandi markets. They are the de-facto kingpins in the coconut marketing system. As channel functionaries, they have been doing a number of marketing functions, and, as channel promoters, their role cannot be underestimated. We have seen two types of wholesalers, one the wholesalers selling within Tamil Nadu and the other, those selling in both inter-state markets and Tamil Nadu. The mean preference score of the respondents towards wholesalers is discussed below:

Table 6.6

Mean preference score of respondents towards wholesalers

Classification of the farmer	*Mean*	*S.D*	*No.*
Small	6.04	2.20	102
Medium	6.39	2.79	102
Big	4.86	3.26	102
TOTAL	5.76	2.85	306

ANOVA

	Sum of Squares	*Df*	*Mean Square*	*F*	*Sig.*
Between Groups	130.824	2	65.412	8.447	**
Within Groups	2346.235	303	7.743		
Total	2477.059	305			

** : Significant at 1% level.

It is observed from the table 6.6 that the mean score of medium farmer is higher (6.39) than the other categories. The calculated F-test value is (8.447) at 1 % level of significance which indicates that there is significant difference among the groups in the average preference towards the wholesalers.

iv) Local Marketers

They are local traders, buying in small quantities directly from coconut farmers for direct sale to consumers. They procure mostly dehusked coconuts directly from marginal and small farmers in villages and local market places on ready cash. They do not employ brokers / agents. They effect sales mostly to ultimate consumers by avoiding unnecessary channel intermediaries which enables them to offer better price to farmers. Coconut farmers possessing coconuts in small quantity that are almost discarded by local marketers in villages and those farmers who do not like to dispose of their coconuts through agents / independent traders, sell their coconuts in local market place itself.

The mean preference score of the sample respondents towards local marketers is discussed under:

Table 6.7

Mean preference score of respondents towards Local Marketers

Classification of the respondents	*Mean*	*S.D*	*No.*
Small	6.83	2.46	102
Medium	6.20	6.82	102
Big	2.50	3.50	102
TOTAL	5.18	5.01	306

ANOVA

	Sum of Squares	*Df*	*Mean Square*	*F*	*Sig.*
Between Groups	1116.725	2	558.363	25.854	**
Within Groups	6543.745	303	21.597		
Total	7660.471	305			

** : Significant at 1% level.

It is noticed from the table 6.7 that the mean score of small farmers respondents is higher than (6.83) medium and big farmer respondents towards local marketers. Small number of production, no overhead cost, and immediate sale are the factors might have motivated the small farmers to have trade relationship with the local marketers. The ANOVA result shows that the calculated F ratio value is 25.854 which is greater than the table value of 4.676 at 1 % level of significance. Thus it can be concluded that there is no significant difference among the farmer groups in the average preference towards the local marketers.

v) Industrial user

Industrial users are also one of the notable non-institutional agencies in coconut marketing. Industrial users have an informal agreement with the coconut farmers for the regular supply of coconuts to their industrial unit. Mostly this source is preferred by the small and medium farmers due to their readiness to offer advance money, farm gate sales, no overhead cost, scope for regular supply, bulk sales etc. Some time industrial users also act as a wholesaler for the neighboring state counterparts. The mean preference scores of the sample respondents towards industrial user are discussed below:

Table 6.8

Mean preference score of Respondent towards Industrial Users

Classification of the farmer	*Mean*	*S.D*	*No.*
Small	6.02	2.57	102
Medium	4.18	6.27	102
Big	3.40	6.54	102
TOTAL	4.53	5.53	306

ANOVA

	Sum of Squares	*Df*	*Mean Square*	*F*	*Sig.*
Between Groups	368.869	2	184.435	6.244	**
Within Groups	8949.304	303	29.536		
Total	9318.173	305			

** : Significant at 1% level

It is clearly understood that the mean score of small farmer respondents (6.02) is higher than medium (4.18) and big (3.40) farmer respondents. The calculated F ratio value is 6.244 which is greater than the table value of 4.676 at 1 % level of significance. Thus it can be concluded that there is significant difference among the farmer groups in the average classification of the farmer in the industrial user.

The table 6.9 clearly pinpoints that non-institutional agencies has played a pivotal role in marketing of coconuts in the study area. The mean score of the quantity and value of coconut marketed by the traders is 40046.4847 and 158508.1081 respectively which occupies first place among the market functionaries. The

Table 6.9

Quantity and value of Coconut Marketed by the Respondents (2005-2006)

(Quantity in number)
(Value in Rupees)

S.No	Channel Members	No. of Respondents	Quantity /Value	Minimum	Maximum	Mean	Std. Deviation
1	Traders	111	Quantity	7200.00	136500.00	40046.4847	24492.4400
		(36.72)	Value	21600.00	492300.00	158508.1081	105339.7181
2	Commission Agents	78	Quantity	5775.00	215500.00	48690.0000	52086.7580
		(25.49)	Value	17325.00	874900.00	205387.8590	224778.5584
3	Industrial User	79	Quantity	3000.00	123480.00	29412.4051	24191.3802
		(5.81)	Value	17325.00	555660.00	118621.1899	106375.6279
4	Itinerant Traders	34	Quantity	5445.00	100348.00	25505.4706	21068.5902
		(11.11)	Value	19057.00	403860.00	100061.2941	85343.0593
5	Cooperatives	4	Quantity	11880.00	92650.00	52265.0000	46632.5812
		(1.30)	Value	41580.00	324275.00	182927.5000	163214.0343

standard deviation score for traders are 24492.4400 and 105339.7181. Whereas the mean score for the quantity and value of coconut marketed by the cooperatives is 52265 and 182927.5 and the standard deviation is 46632.5812 and 163214.1343.

Coconut Growers' Perception about the Present Marketing Functionaries

In this section, an in-depth analysis is made on the following factors enabling for assess the coconut grower's perception about the present market functionaries in three geographical zones viz. high, moderate and low coconut production within the study area.

Marketing of coconut, pattern of disposal, sale practice, opinion about existing system of marketing, problems in marketing of coconut, quantity and value of coconut marketed, average market price, satisfaction over the channel member, suggestion to overcome marketing problems, reasons for preferring institutional and non-institutional agencies, role of institutional agencies, problems with the institutional agencies, advantages in the institutional agencies, role of non-institutional agencies and its merits and demerits, marketing practices followed by the non-institutional agencies, perception about the present market functionaries both institutional and non-institutional agencies etc., are considered for statistical analysis.

The researcher desires to be acquainted with the perception of the coconut growers (N=306) in relation to the present market functionaries operating in the study area. For this insight analysis a set of questionnaires has been included in the schedule itself. The respondents are asked to mark their views in the appropriate column provided for each of the market functionaries. The questions regarding their membership, trade practices, opinion towards price, standards, service, grading, problems, facilities, cost of sale, loan facilities, market information etc. were asked for the purpose of coming to a conclusion about their acuity towards the present functionaries.

Table 6.10: Sources of Technical Knowhow

S. No.	*Source of Technical know how*	*No. of Respondents*	*Percentage*
1	None	8	2.6
2	Government Institutions	121	39.5
3	Experienced farmers	172	56.3
4	Traders	5	1.6
	Total	306	100.0

Always professional knowledge fetches best result in all fields. There is no exception for coconut production, it is very essential to have adequate experience and professional knowledge to witness fruitful result during the harvest season. It is clearly evident from the study that majority of the sample farmers (56.3 %) acquainted with the technical aspects of coconut production from the experienced farmers followed by Government institutions (39.5%). However, the role played by the traders in this regard found to be insignificant.

Table 6.11
Choice of Channel for Trading

S.No.	*Market Functionary*	*No. of Respondents*	*Percentage*
1	Cooperative Marketing Society	84	27.4
2	Regulated Market	01	0.4
3	Traders	101	33.1
4	Commission Agents	43	14.0
5	Wholesalers	42	13.7
6	Local Marketers	16	5.2
7	Industrial User	19	6.2
	Total	306	100

A simple majority of the respondents (33.1%) chosen the traders as their channel member followed by cooperative marketing society (27.4 %). Flexibility in trade operations, advance payment, farm gate purchase, less cumbersome trade process might have influenced the farmers to go behind the traders. Standardized practice, warehousing facility, advance against the pledge, reasonable price, open and free sale process might have attracted the farmers to have trade relationship with the cooperative marketing society. Nevertheless, rest of the non-institutional agencies like commission agents, wholesalers, industrial user and local marketers also attracted the farmers in a notable way.

Table 6.12: Relationship between Marketing Channels and Price Offered by them

Marketing Channel	*Response*						*TOTAL*	
	Satisfied		*Not satisfied*		*Neither satisfied nor dissatisfied*			
	No.	*%*	*No.*	*%*	*No.*	*%*	*No.*	*%*
Traders	67	60.4	38	34.2	6	5.4	111	100.0
Commission Agents	42	53.8	28	35.9	8	10.3	78	100.0

Industrial user	35	44.3	32	40.5	12	15.2	79	100.0
Cooperatives	18	52.9	8	23.5	8	23.5	34	100.0
Local Marketers			2	50.0	2	50.0	4	100.0
TOTAL	162	52.9	108	35.3	36	11.8	306	100.0

Chi-Square Test

	Value	df	Sig.
Chi-Square	20.352	8	**

**: Significant at 1% level.

It is clearly observed from the table that majority of the respondents (52.9) have satisfied the price offered by their marketing channels. 60.4, 53.8, 44.3 and 52.9 percent of the respondents have satisfied the price offered by traders, commission agents, industrial users and cooperatives respectively. None of the respondents has satisfied with the local marketers.

Chi square test was applied to find whether there is significant relationship between marketing channels and price offered by them. The calculated value of chi square is 20.352 which are greater than the table value of 20.090 at 1% level of significance. Since the calculated value is greater than the tale value it is inferred that there is significant relationship between the marketing channels and price offered by them. Hence HO_2 : "There is no significant relationship between marketing channels and price offered by them" is rejected.

Table 6.13

Good Marketing Practices followed by

S. No.	*Market Functionary*	*No. of Respondents*	*Percentage*
1	Cooperative Marketing Society	62	20.1
2	Regulated Market	02	0.7
3	Traders	91	29.8
4	Commission Agents	59	19.4
5	Wholesalers	41	13.4
6	Local Marketers	24	7.8
7	Industrial User	27	8.8
	Total	306	100

Good marketing practices like exact counting, using correct weighing machine, timely payment, price according to the quality, ethical practice, open and free trading will give utmost satisfaction to the coconut producers. A considerable majority of the respondents (29.8 percent) opined that the above said practices are followed by the traders. Next to the traders Cooperative Marketing Society occupies a place in the hearts of the sample farmers (20.1 percent). None the less Commission agents (19.4 percent) and wholesalers (13.4 percent) are also following good marketing practices in their marketing operations.

Table 6.14

Reasonable Grading is Done By

S.No.	*Market Functionary*	*No. of Respondents*	*Percentage*
1	Cooperative Marketing Society	26	8.5
2	Traders	59	19.3
3	Commission Agents	52	17.0
4	Wholesalers	85	27.8
5	Local Marketers	38	12.4
6	Industrial User	46	15.0
	Total	306	100

Grading is necessary in all sorts of business especially in agricultural business. Grading acts as a yardstick to fix the price for the products. Usually, according to the grade / quality of the produce the price will be fixed by the market functionary. In this grading activity, a good majority of the respondents opined that the wholesalers do justifiable job (27.8 percent). Rest of the sample respondents favoured their opinion towards reasonable grading to the following market functionary's viz., traders (19.3 percent), commission agents (17 percent), industrial user (15 percent) and cooperative marketing society (8.5 percent).

Table 6.15

Service Oriented Functionary

S. No.	*Market Functionary*	*No. of Respondents*	*Percentage*
1	Cooperative Marketing Society	186	60.8
2	Regulated Market	16	5.2
3	Traders	28	9.2
4	Commission Agents	25	8.2
5	Wholesalers	17	5.6
6	Local Marketers	23	7.5
7	Industrial User	11	3.5
	Total	306	100

Perusal of the Table 6.15 revealed that majority (60.8 %) of the respondents found Cooperative Marketing Society as the best service oriented marketing functionary. It is distressing to note that Regulated Market being the government run institution didn't make much headway as the response rate among the respondents was found to be insignificant. Nevertheless, 34 percent of the respondents gave due consideration on the service orientation of non-institutional market functionaries.

Table 6.16

Likely to Continue Trade With

S.No.	*Market Functionary*	*No. of Respondents*	*Percentage*
1	Cooperative Marketing Society	44	14.4
2	Regulated Market	02	0.7
3	Traders	67	21.9
4	Commission Agents	52	17.0
5	Wholesalers	43	14.1
6	Local Marketers	38	12.4
7	Industrial User	60	19.5
	Total	306	100

Inspite of different opinion among the farmers about the non-institutional agencies, they like to continue trade with them only as it was witnessed from the table 3 that as high as 84.5 % of the respondents desired to continue trade with them. This might be due to the fact that non-institutional functionaries as compared to institutional functionaries seemed to adopt high flexibility in terms of rule, procedures and regulations.

Table 6.17

Problematic Market Functionary

S. No	*Market Functionary*	*No. of Respondents*	*Percentage*
1	Cooperative Marketing Society	30	9.8
2	Regulated Market	06	1.9
3	Traders	35	11.4
4	Commission Agents	99	32.4
5	Wholesalers	51	16.7
6	Local Marketers	49	16.0
7	Industrial User	36	11.8
	Total	306	100

Though the non-institutional agencies hold giant share in

the volume of business with the coconut farmers, they are in the problematic side. Most of the respondents (32.4 percent) felt that commission agents are the most problematic in the market. Only 9.8 percent of the sample respondents inferred that cooperative marketing society is also one of the problematic market functionaries in the coconut market.

Table 6.18

Timely Payment Made By

S. No.	*Market Functionary*	*No. of Respondents*	*Percentage*
1	Cooperative Marketing Society	25	8.2
2	Regulated Market	01	0.4
3	Traders	69	22.5
4	Commission Agents	48	15.7
5	Wholesalers	44	14.4
6	Local Marketers	64	20.9
7	Industrial User	55	17.9
	Total	306	100

In agricultural marketing timely payment is the deciding as well as motivating factor which leads the farmers to have continuous trade relationship with market functionaries. The survey result in this regard revealed that majority of the respondents opined that timely payment is being effected by traders, local marketers, industrial user and commission agents.

Table 6.19

Source of Short-Term Loans

S.No.	*Market Functionary*	*No. of Respondents*	*Percentage*
1	Cooperative Marketing Society	20	6.5
2	Regulated Market	02	0.7
3	Traders	51	16.7
4	Commission Agents	62	20.3
5	Wholesalers	53	17.3
6	Local Marketers	67	21.8
7	Industrial User	51	16.7
	Total	306	100

Short-term loans are very significant for the farmers to meet the day-to-day expenses in the agricultural field. Most of the farmers seek this type of loans from the informal sector because of its less cumbersome procedure. This is proved in this research since the survey result showed that majority of the short- term financial needs of farmers are being met out by the non-institutional agencies (92.8 percent).

Table 6.20
Source of Long-Term Loans

S.No.	Market Functionary	No. of Respondents	Percentage
1	Cooperative Marketing Society	12	3.8
2	Regulated Market	10	3.6
3	Traders	48	15.7
4	Commission Agents	42	13.6
5	Wholesalers	93	30.3
6	Local Marketers	49	16
7	Industrial User	52	17
	Total	306	100

Long-term loans are very essential for the farmers those who are growing perennial crops because the yielding takes place only after a long period. It is observed that majority of the farmers (30.3 percent) have obtained long-term loans from the wholesalers. Industrial user (17 percent), local marketers (16 percent), traders (15.7 percent) also played a considerable role in issuing long-term loans to the coconut growers.

Table 6.21
Good Pre-procurement Services

S.No.	Market Functionary	No. of Respondents	Percentage
1	Cooperative Marketing Society	12	3.8
2	Regulated Market	06	1.9
3	Traders	59	19.3
4	Commission Agents	61	20.0
5	Wholesalers	61	20.0
6	Local Marketers	75	24.5
7	Industrial User	32	10.5
	Total	306	100

Pre-procurement services are very common today because of heavy competition in every market. Coconut markets also have no exception in providing the pre-procurement services to the farmers. Generally, pre-procurement services in coconut marketing includes advance payment, provision of fertilizers, short-term credit, long term credit provision of godown facilities, transport facilities, loading and unloading etc. The survey result shows that non-institutional agencies (94.3%) have been playing a significant role in pre-procurement process.

Table 6.22
Good Post-procurement Services

S.No.	*Market Functionary*	*No. of Respondents*	*Percentage*
1	Cooperative Marketing Society	07	2.3
2	Regulated Market	03	0.9
3	Traders	78	25.5
4	Commission Agents	40	13.1
5	Wholesalers	60	19.6
6	Local Marketers	71	24.2
7	Industrial User	44	14.4
	Total	306	100

Today in agricultural marketing there is a dire need for post procurement service to the farmers for the continuous trade relationship with the marketing functionaries. It was observed that 25.5 percent of the respondents have high opinion towards traders in this regard may be due to the facts like supply of fertilizer at the door steps, provision of finance in advance, personal intimacy etc. On the other hand local marketers also play a considerable role (24.2) in providing post sales service to the coconut growers.

Table 6.23
High Malpractices

S.No.	*Market Functionary*	*No. of Respondents*	*Percentage*
1	Cooperative Marketing Society	24	7.9
2	Regulated Market	02	0.7
3	Traders	62	20.2
4	Commission Agents	106	34.7
5	Wholesalers	31	10.1
6	Local Marketers	45	14.7
7	Industrial User	36	11.7
	Total	306	100

Malpractices are part and parcel of agricultural marketing. Coconut marketing is also not exception in this regard. It is clearly understood that 34.7 percent of the respondents opined that commission agents do various malpractices during their marketing operations followed by 20.2 percent traders. Malpractices in Cooperative Marketing Society are found to be low as compared to other marketing functionaries.

Table 6.24
Easy Accessibility

S. No	*Market Functionary*	*No. of Respondents*	*Percentage*
1	Cooperative Marketing Society	22	7.2
2	Regulated Market	08	2.6
3	Traders	50	16.4
4	Commission Agents	73	23.8
5	Wholesalers	57	18.6
6	Local Marketers	63	20.6
7	Industrial User	33	10.8
	Total	306	100

Transporting the produces from the production point to the market place is a Herculean task to the farmers because it needs human resource, time as well as transportation facility. To overcome such problems the farmers interested to sell their products to the nearby market functionary. It is observed from the above table that majority of the respondents (23.8 percent) opined that commission agents are the only nearby source to market their produces. 20.6 and 18.6 percent of the respondents viewed their opinion towards local marketers and wholesalers respectively.

Table 6.25
Hassle Free Trade

S. No.	*Market Functionary*	*No. of Respondents*	*Percentage*
1	Cooperative Marketing Society	12	3.8
2	Regulated Market	10	3.6
3	Traders	74	24.1
4	Commission Agents	57	18.6
5	Wholesalers	63	20.6
6	Local Marketers	44	14.3
7	Industrial User	46	15
	Total	306	100

Now days the farmers expect less cumbersome trade practices in their marketing operations. It is evident that most of the farmers (24.1 percent) felt that trader follows hassle free trade and followed by wholesalers (20.6 percent). Rest of the respondents felt that commission agents (18.6 percent), industrial user (15 percent) and local marketers (14.3 percent) follow less cumbersome process in their marketing activities. Only 3.8 percent of the sample respondents opined that cooperatives also follow hassle free trade in their marketing operations.

Table 6.26
Godown Facilities

S.No.	*Market Functionary*	*No. of Respondents*	*Percentage*
1	Cooperative Marketing Society	75	24.5
2	Regulated Market	20	6.5
3	Traders	38	12.4
4	Commission Agents	24	7.9
5	Wholesalers	59	19.3
6	Local Marketers	52	17.0
7	Industrial User	38	12.4
	Total	306	100

During the flush season the coconut growers may not be in a position to withhold their produces for a long time because of less storage facilities in their houses. So, the farmers are in need of godown facilities to store their produces. It is clearly said in the table (6.45) that majority of the respondents (24.5 percent) felt that cooperative marketing society provides godown facilities for their produces. Wholesalers (19.3 percent), local marketers (17 percent) and industrial user (12.4 percent) are also kind enough to extend their space for the storage purpose of the farmers.

Table 6.27
Provision of Market Information

S.No.	*Market Functionary*	*No. of Respondents*	*Percentage*
1	Cooperative Marketing Society	25	8.3
2	Regulated Market	06	1.9
3	Traders	66	21.7
4	Commission Agents	41	13.4
5	Wholesalers	69	22.5
6	Local Marketers	60	19.5
7	Industrial User	39	12.7
	Total	306	100

Frequent market information is necessary for the farmers to know the present price status of each and every coconut products in the market. Most of the farmers (22.5 percent) felt that wholesalers provide enough market information about the market. Next to the wholesalers traders stood second (21.7 percent) in providing relevant market information to the farmers. Local marketers (19.5 percent), commission agents (13.4 percent) and industrial user (12.7 percent) also plays a considerable role in providing market information. Only 8.3 percent of the respondents said that cooperative marketing societies also provide some market information to the respondents.

Table 6.28

High Price fluctuations Prevail In

S.No.	*Market Functionary*	*No. of Respondents*	*Percentage*
1	Cooperative Marketing Society	20	6.5
2	Regulated Market	06	1.9
3	Traders	40	13.1
4	Commission Agents	102	33.4
5	Wholesalers	52	17.0
6	Local Marketers	52	17.0
7	Industrial User	34	11.1
	Total	306	100

Generally high price fluctuations prevail in agricultural marketing. It is the part and parcel of each and every agricultural product. Coconut product also has no exception in high price fluctuation. But wide price fluctuations generally prevail only in non-institutional agencies. Among the non-institutional agencies commission agents have high price fluctuation in their marketing operations (33.4 percent). Next to commission agents, the respondents felt that there are high price fluctuations among the wholesalers (17 percent) and local marketers (17 percent). Only 6.5 percent of the respondents felt that there is a high price fluctuation in cooperative marketing society.

Table 6.29

Lack of Standardized Marketing Practice

S.No.	*Market Functionary*	*No. of Respondents*	*Percentage*
1	Cooperative Marketing Society	27	8.8
2	Regulated Market	04	1.3
3	Traders	65	21.3
4	Commission Agents	95	31.0
5	Wholesalers	51	16.7
6	Local Marketers	39	12.7
7	Industrial User	25	8.2
	Total	306	100

In market, standardized practice is necessary to each and every market functionary for their growth, development and excel. But standardized practice lacks generally in agricultural marketing and especially in coconut marketing. Majority of the respondents (31 percent) opined that commission agents lack some standardized practice in their marketing operations. Next to commission agents, traders stood second (21.3 percent) in lacking standardized practice. Rest of the sample population opined that wholesalers (16.7 percent), local marketers (12.7 percent) and industrial user (8.2 percent) lacks in practicing standards in their marketing operations. Only 8.8 percent of the respondents opined that cooperative marketing society also lacks standardized practice in their marketing operations.

Table 6.30
Value Addition Services

S.No.	*Market Functionary*	*No. of Respondents*	*Percentage*
1	Cooperative Marketing Society	30	9.9
2	Regulated Market	05	1.6
3	Traders	54	17.5
4	Commission Agents	58	19
5	Wholesalers	48	15.7
6	Local Marketers	46	15
7	Industrial User	65	21.3
	Total	306	100

Value addition is the process of face lifting the raw coconut into semi-useful form. Value addition services are not a common practice followed by the entire market functionary. This type of exclusive function only is handled by industrial user. This is the reason that most of the respondents (21.3 percent) opined that value addition services are more in the industrial user. Commission agents also took part considerably in value addition activities (19 percent). A good majority of the respondents (17.5 percent) gave their consent towards traders in this value addition process. Out of 306 samples only 30 sample populations (9.9 percent) viewed that cooperative marketing society also undertakes value addition activity.

Table 6.31
High Cost of Sale

S.No.	Market Functionary	No. of Respondents	Percentage
1	Cooperative Marketing Society	120	39.3
2	Regulated Market	04	1.3
3	Traders	43	14.0
4	Commission Agents	44	14.4
5	Wholesalers	53	17.3
6	Local Marketers	27	8.8
7	Industrial User	15	4.9
	Total	306	100

Cost of sale refers that amount spent for transportation, loading, unloading, packing etc from the production point to the market yard. Now-a-days, the cost of sale is high in every sector especially in agricultural sector due to paucity of human resource. From the above analysis table, it is cleared that majority of the respondents (39.3 percent) said that the cost of sale is high in cooperative marketing society. This is may be because of the institutional agencies like cooperatives and regulated market cannot adopt the practices followed by the non-institutional agencies like free transportation, farm gate pick up etc. 17.3 percent of the respondents opined that there is a high cost of sale they face while marketing their produces to the wholesalers. Rest of the respondents opined that there is a high cost of sale among the commission agents (14.4 percent) and traders (14 percent).

Table 6.32
Issue of Pledge Loans

S.No.	Market Functionary	No. of Respondents	Percentage
1	Cooperative Marketing Society	83	27.1
2	Regulated Market	09	2.9
3	Traders	38	12.4
4	Commission Agents	61	19.9
5	Wholesalers	50	16.4
6	Local Marketers	34	11.1
7	Industrial User	31	10.2
	Total	306	100

Pledge loans are mostly availed by the farmers when the market is not conducive to sell the produces. Generally these types of loans are available only with the institutional agencies and very rarely it is available with the non-institutional agencies. It is inferred from the above table that majority of the respondents (27.1 percent) availed pledge loans from the cooperative marketing society. It is observed that commission agents played a considerable role (19.9 percent) in issuing pledge loan to the coconut peasant community. It is notable that the Wholesalers (16.4 percent) and traders (12.4 percent) also have a practice of issuing pledge loans to the farmers.

Table 6.33

Farm Gate Purchase

S.No.	*Market Functionary*	*No. of Respondents*	*Percentage*
1	Cooperative Marketing Society	14	4.6
2	Regulated Market	07	2.3
3	Traders	60	19.6
4	Commission Agents	70	22.9
5	Wholesalers	60	19.6
6	Local Marketers	60	19.6
7	Industrial User	35	11.4
	Total	306	100

Now-a-days farm gate purchase is undertaken by almost all the market functionaries except institutional agencies like cooperative marketing society and regulated market. The farmers also very much interested to have business dealing only those who have undertaking farm gate purchase. This may be because of minimizing the cost of marketing especially transport cost. The above table clearly shows that most of the sample respondents accepted that the non-institutional agencies (93.1 percent) offered the farm gate purchase. Only 6.9 percent of the respondents said that institutional agencies also do this type of arrangements to the coconut farmers.

Table 6.34

Most Advantageous Functionary

S.No.	*Market Functionary*	*No. of Respondents*	*Percentage*
1	Cooperative Marketing Society	18	5.8
2	Regulated Market	10	3.6
3	Traders	71	23.3

4	Commission Agents	65	21.2
5	Wholesalers	38	12.3
6	Local Marketers	57	18.5
7	Industrial User	47	15.3
	Total	306	100

Most advantageous refers that the farmer should have a full benefit in all respects over his marketing operations with a particular market functionary. It is inferred from the above table that a considerable majority of the respondents (23.3 percent) opined that the traders are the most advantageous market functionary in the market. Next to the trader, it is said that commission agents (21.2 percent), local marketers (18.5 percent), industrial user (15.3 percent) and wholesalers (12.3 percent) are also in a position to give some benefit to the farmers. Only 5.8 percent of the respondents opined that cooperative marketing society is one of the most advantageous market functionaries in the market.

Table 6.35
High Technical Know How

S.No.	*Market Functionary*	*No. of Respondents*	*Percentage*
1	Cooperative Marketing Society	10	3.6
2	Regulated Market	00	00
3	Traders	73	23.7
4	Commission Agents	68	22.2
5	Wholesalers	52	17.0
6	Local Marketers	54	17.5
7	Industrial User	49	16.0
	Total	306	100

Technical know high refers that possessing high knowledge in local as well as international coconut market. From the above table it is known that majority of the respondents (23.7 percent) opined that the traders have high technical knowledge among the market functionaries. 22.2 percent of the samples have an opinion that commission agents also possess technical know how about the coconut market. Only 3.6 percent of the respondents favoured towards cooperative marketing society in this regard.

Table 6.36

Open and Free Sale Process

S.No.	*Market Functionary*	*No. of Respondents*	*Percentage*
1	Cooperative Marketing Society	17	5.5
2	Regulated Market	04	1.3
3	Traders	82	26.7
4	Commission Agents	64	21.0
5	Wholesalers	51	16.7
6	Local Marketers	37	12.1
7	Industrial User	51	16.7
	Total	306	100

Now-a-days open and free sale process is a difficult one among all the market functionaries in the market. A good majority of the sample respondents (26.7 percent) opined that traders follow somewhat open and free sale proceeds in their marketing operation. 21 percent of the respondents said that commission agents also follow this type of open and free marketing operations. Only 5.5 percent of the sample respondents favoured cooperative marketing society in this regard.

Table 6.37

Overall Service

S.No.	*Market Functionary*	*No. of Respondents*	*Percentage*
1	Cooperative Marketing Society	56	18.3
2	Regulated Market	00	00
3	Traders	136	44.5
4	Commission Agents	45	14.7
5	Wholesalers	23	7.5
6	Local Marketers	20	6.5
7	Industrial User	26	8.5
	Total	306	100

Overall service refers to the services rendered by the market functionaries from the farm gate purchase to till the payment is made. It is clear from the above table that a good majority of the sample respondents (44.5 percent) opined that the overall service is good among the traders. Next to traders, a good majority of the respondents (18.3 percent) inferred that cooperative marketing society is also performing good service to the peasant community. 14.7 percent of the respondents favoured their opinion in this regard to the commission agents. Rest of the 8.5, 7.5 and 6.5 percent of the respondents exercised their right towards industrial user, wholesalers and local marketers.

Sum up

To sum up the Indian Coconut cultivation possesses a very bright future despite the constraints and problems beset with the industry in certain areas. Available infrastructure, trained manpower and wide range of climatic conditions in the country are indicative of best capabilities. At the same time, these resources have to be effectively utilized to harness the best for making coconut industry more competitive and dynamic. India will continue to be the premier producer in the world and also assume leadership in coconut trade.

7

Findings, Conclusions and Suggestions

The coconut palm is essentially a tropical crop and is grown in countries located in the tropical region between latitudes 22° N-27°S. The distribution of palm extends over most of the island, peninsular and coastal areas and even in some places outside the tropic zone. The main coconut growing areas in the world are located in Asia, Oceania, the West Indies, Central and South America and West and East Africa. In Asia, it is rarely found in the north of India. In India, the palm grows as far north as Lucknow. In Africa, the northern limits are Cápe Verde (15° N) on the West Coast and Djibouti (11° 30″N) on the East, while the Southern limits are Mossamedas (15° S) on the West Coast and the Zambesi river (19° S) on the East. In the Pacific, the palm is found as far north as the Bonin Islands (26° N) and as far South as the Pitcairn Islands (25° S). In South America the palm is found in Brazil at latitude of 27° S and in North America in Florida and Bahamas at latitude of 25° N and in Mexico. It is grown in nearly ninety countries located in the tropical belt of the world.

The coconut palm is the most extensively grown nut in more than 90 countries in the world. It is grown in an area of about 12 million hectares in the world, which account for about 55.569 billion nuts. More than 84 percent of the production is from Asia. The Indonesia, Philippines and India altogether account for more than 76.18 percent and 75.73 percent in the area under coconut and the production of nuts respectively. The coconut palm is essentially a crop of small farmers and is, perhaps, the only source of livelihood to millions of farmers in almost all the countries where it is grown. The coconut industry provides livelihood for one million, three million, three million and 10 million people living in Malaysia, Indonesia, Sri Lanka and India respectively. It supports more than 80 million small and marginal farmers in the world. Indonesia, Philippines, India and Sri Lanka are the major players in the global coconut scenario, and in all these countries the coconut exerts profound influence in the economy. In India, coconut is not only associated with rural economy but is interwoven in cultural heritage.

Among coconut growing countries, Asia and the Pacific regions produce 48448 million nuts from 10.65 million ha sharing 87.19 percent in production and 90.26 percent in area. India, Indonesia, the Philippines and Sri Lanka are the four major coconut growing countries, which together contribute 80.73 percent of the world's production.

During the period 1990-2005, there has been 5.64 per cent increase in area, 17.41 per cent increase in production but only 11.15 per cent increase in productivity. However, production continued to increase till 1995 and then there was a sharp decline both in production and productivity up to 2000. Declining trend in productivity after 1995 could be attributed to increasing number of old and senile trees. But the year 2000 onwards there was a steep increase both in production and productivity due to the high yield varieties.

Indonesia ranks first in coconut production and area by contributing 14,984 million nuts from 3.81 million ha., sharing 32.31 per cent in area and 26.96 per cent in production. Coconut in Indonesia is grown mainly by small holders and forms a major economic activity of about 2.5 million farm families. In Philippines, coconut is grown in 3.24 million ha producing 12,600 million nuts, ranking second in area and third in production. A coconut industry support nearly one-third of Philippines' population and it is basically a small holders' crop. The percent share of Philippines in area and production is 27.49 m ha. and 22.68 m. nuts respectively. In India, the area and production of coconut are 1.93 million ha. and 12,600 million nuts respectively. The crop supports 11 million people, mainly living in the coastal tracts in the country. It contributes Rs.72, 000 million to the Gross Domestic Product and Rs.3,620 million to the foreign exchange earnings mainly from export of coir products

The other countries, which contribute significantly to the world production of coconut, are Bangladesh, Papua New Guinea and Myanmar, smaller islands viz., Vanuatu, Solomon Islands, Samoa, Fiji, Palau and Micronesia. Although these countries have small share in coconut production, the economies of these countries are significantly benefited by coconut.

There is distinct difference in the pattern of distribution of the coconut in the country. Bulk of the coconut production is contributed by the four southern states viz., Kerala, Karnataka, Tamil Nadu and Andhra Pradesh. Other traditional coconuts growing States/Union Territories (UT) in the country are Orissa, Maharashtra, West Bengal, Goa, Lakshadweep and Andaman and Nicobar Islands. The crop now is grown in favorable locations in the Central, North and North-Eastern regions of the country. Of the present area of 1.90 million ha in the country, the contribution of Kerala in area is 0.82 million ha. followed by Karnataka (0.40 million .ha.), Tamil Nadu (0.38 million ha.) and Andhra Pradesh (0.1 million ha.) sharing 43.02 per cent,21.28 per cent,20.14 per cent and 5.3 per cent respectively, which together accounts for 89.7 per cent of the total area under coconut in the country. In the production of coconut, Kerala tops the list with 5641 million nuts with a share of 38.26 per cent followed by Tamil Nadu (33.70 %), Karnataka (11.08 %) and Andhra Pradesh (7.59 %).

The country has recorded a sizeable increase in area, production and productivity during the last one decade. The area increased from 1.51 million ha to 1.90 million ha. with an increase of 0.39 million ha and production from 9,700 million nuts to 14,743 million nuts with an increase of 5,043 million nuts. The productivity increased from 6,407 nuts to 7747 nuts per ha. Thus it could be seen that despite the undulating growth behavior exhibited in production by a few states, the Indian scenario presents a healthy trend in expansion in area and increase in production for the last few years.

The Tamil Nadu state recorded an increase in cultivated area under coconut from 0.66 million ha during 1950-51 to 383 million ha during 2007-2008, during the 50 year period. The share of the State in the area, under coconut has gone up from 10.54 per cent of the national area during 1950-51 to 20.14 per cent during 2007-2008 (CDB, 2001). This would mean that the increase in the area under cultivation has been going up at a more rapid rate than it was in many other states including Kerala and the National average.

The production of coconut in Tamil Nadu has gone up from 462 million nuts during 1950-51 to 4968.20 million nuts during 2007-2008 registering a 1073 per cent of increase during the 57 year period. The share of Tamil Nadu's production to national coconut production has gone up from 14.08 per cent during 1950-51 to 33.70 per cent during 2007-2008. The increase in production in the State had been far above the national average increase (CDB, 2003). Tamil Nadu has also made impressive gains in productivity. The productivity of coconut in the state has gone up from 7,009 nuts per ha during 1950-51 to 12,959 nuts per ha during 2007-2008 which is far above the national average of 7,608 nuts per ha.

In spite of the excellent investment opportunities offered by the crop, Tamil Nadu is lagging behind in the field of production of diversified coconut products and coconut based agro industries. Production of coconut products particularly non-conventional coconut products such as coconut milk and powder, coconut cream, coconut gel, vinegar, processed tender nut water, coconut chips etc., are yet to be initiated. As of now, the coconut processing sector in Tamil Nadu is largely represented by coconut oil mills / industries and a few coconut desiccated units and the non-kernel based coconut industries such as coconut fibre and coir units. The other by-products such as timber, leaves, midribbes etc., are mal used, although they can be converted into value added products having high export potential.

Marketing is as important as production to any producer. Because it creates value to the product; it pays revenue to the producer; and more than these, it directs the producer as to whether continue or stop production. To a farmer, marketing is something more than production, due to certain inherent features that neither the production can be controlled/ regulated in tune to market changes due to predominance of natural forces affecting production functions nor can marketing be performed in tune to market requirements due to his own internal constraints. Hence farmers remain with chronic problem that they can neither derive the advantages of their production/ productivity increase nor the advantages of better marketing. In all the years farmers were deprived of the benefits of increased production in terms of market price and vice versa. Equalization of demand and supply functions in agricultural marketing is far from anybodies guess. Coconut farmers are not exempted from the above phenomenon.

Marketing determines the productivity and profitability. Creating a sound marketing system is essential, as the problems of coconut growers mostly emerge due to the imperfections in the marketing system. Sincere attempts have been made to clear-off the marketing imperfections especially after independence during plan periods. Development of different forms of marketing institutions questions the relevance of the conventional formal marketing structures. Non- institutional agencies still dominate in coconut marketing. If marketing is imbibed with imperfections, coconut farmers should have received low returns. This in turn could have resulted for reduction in coconut production. On the contrary, it is amazing to notice that both the area under coconut cultivation and production of coconut have been increasing year after year. The foregoing issue motivates the researcher to conduct a new research study in the sample area and to find out the trends in coconut production, marketing practices of coconut growers, role of institutional and non-institutional agencies in coconut marketing and perception of the coconut growers about the present market functionaries.

The principal objectives of the present study are to analyze the production practices of coconut growers, to assess the marketing practices of coconut marketing channels and to evaluate the perception of coconut growers about the present scenario of coconut marketing.

Multistage sampling procedure was adopted for the present study. 51 farmers each from six villages totally 306 farmers were selected for the study by using quota sampling procedure.

Major Findings

The important findings and conclusions drawn from the analysis done in all chapters are summarised hereunder:

Production practices of coconut growers

The majority of the respondents are males and nearly half of them are found to be old age category. It is quite appreciable that all the respondents had formal education. As to family size, majority of the respondents had medium size family which might be due to the prevalence of literacy among coconut growers. All the respondents possessed more than 10 years experience in their occupation. Majority of the respondents are cultivating coconut on their own land. Lease and tenant coconut were also found among less proportion of coconut growers. A good majority (74.5%) of the respondent average annual income was worked out to be less than 2 lakhs. Coconut is found to be the major source of income of the respondents. Coconut traders, cooperative banks and commercial banks have been playing significant role in extending credit facilities to the coconut farmers.

A sizeable portion (43.8 %) of the respondents clearly viewed that marginal labour force is the motivating factor for the cultivation of coconut followed by 38.6 percent and 17 percent of the respondents expressed that profitability and perennial are the motivational factors for coconut cultivation respectively.

It was found in the survey that majority (43.8 percent) of the respondents preferred grove variety followed by 36.6 percent of the respondents opted peripheral variety. Nonetheless 19.6 percent of farmers opted for both the varieties to fetch the maximum benefits.

The result of the chi square test shows that there is significance relationship between educational qualifications of the respondents and variety of coconut cultivation.

A vast majority of the respondents irrespective of their age group (85.6 percent) opined that coconut cultivation is better than other crops. This may be due to the fact that profitability, perennial character etc while 14.4 percent of the respondents having negative opinion. This may be because of their bitter experience with the production and marketing aspects. The chi square value also confirms that (2.422) there is no significant relationship between the age group and their response about coconut cultivation.

Majority of the respondents, irrespective of their classification (57.5 percent) were not interested to extend their area under coconut, owing to the fact that non-availability of vacant land, less water source, inability to manage the farm etc., and rest are interested to extend their area under coconut cultivation. The chi square value confirms that there is a significant relationship between the size of land holdings and their interest towards extending area under coconut.

A good majority of the respondents irrespective of their agricultural land status (73.9 percent) depends on only coconut farming. This may be due to the fact that to reap maximum benefit from the soil as well to enhance the quality of the products and to keep intact the quality of the soil for a longer period. The result of chi square test shows that there is no significant relationship between agricultural land status and dependency over coconut trees.

It is very perceptible from that there is a moderate demand both for male (64.1 percent) and female labour (41.2 percent). It was also confirmed with the result of informal discussion held with the cultivators that there is a heavy demand for semi-skilled workers like tree climbers and husk removers.

A good majority of the respondents irrespective of their level of income (61.4 %) preferring to reap in the form of coconut rather than copra and tender coconut. The reason attributed for this state are hassle free harvest, permanent demand for coconut, profitability etc. It is obvious to note that almost all the low income category farmers are reaping in the form of tender coconut may be due to meet their unexpected expenses, regular flow of income to the family etc.

It is apparent that majority of the eco-friendly sample respondents (62.4%) are following organic farming irrespective of its cost. Nonetheless 37.6 percent of the respondents still following inorganic farming owing to its cost effectiveness, readiness, easy availability, subsidy from the government etc.

An attempt was made through chi square test to find significant relationship with two variables namely years of experience in agricultural business and type of input used. The result of the test shows that there is significant relationship between the variables.

A good number of farmers (176, 57.5%) opted the private agencies for their saplings followed by 117 (38.2%) chosen the Government departments. It is a universal awareness that private agencies are highly professional than that of Government sector. Moreover there is a pre and post sales services for the purchasers. It is appreciable that a very meagre number of farmers are using their own indigenous seed for their cultivation.

It is clear that a good majority (44.4 percent) of the respondents chosen hybrid varieties than that of ordinary variety in their cultivable land. It is interesting to point out that young age group respondents (84%) chosen the hybrid variety massively. It is further confirmed by the chi square test as the calculated chi square value is higher (60.799) than the table value, it is inferred that there is a significant relationship between the age of the respondents and the variety of the tree planted.

The minimum number of trees planted per acre is same for both the varieties i.e., 60 but the maximum has a meagre difference. The minimum age of the periphery and grove tree is 9 and 4 years and maximum for both the variety is 40 years. The mean age of the periphery tree is 22.4767. The mean age of grove tree is 10.7526 The minimum and maximum years taken to give yield by a periphery and grove tree is 5 and 9 years and 4 and 7 years respectively. The sample farmers in the study area follows uniform pattern of distance between the two trees. The mean annual productivity of the periphery and grove tree is 113 and 151 respectively.

To know the significant difference between the grove and peripheral variety trees in the average productivity per tree t - test and paired sample t-test was applied. The result of the test shows that there is a significant difference between the two varieties.

One way ANOVA test was applied to find whether there is significant difference among the grove and peripheral coconut groups in the average total sales per acre. The ANOVA result shows there is significant difference among the variety of coconut groups in the average of total sales value per acre.

In study area majority of respondents are undertaking production enhancement operations such as crop pruning, rope support, inter ploughing, inter cropping which enabled the farmers in terms of productivity increase, product quality, life span enhancement etc.

Majority (46.4 percent) of the respondents opined that they are badly in need of medium term loans for the purpose of meeting out the recurring expenses of the firm. It is appealing to note that 3.9 percent of the respondents have the financial capacity to look after the firm by their own wealth. The calculated value of chi square is 29.608 which is higher than the table value of 16.812 at 1% level of significance which confirms that there is significant relationship between these two variables.

A vast majority of the respondents (78.8%) have purchased inputs from PACBs. Indeed it is common knowledge that less price, good quality, subsidy from the government, nearness to the village, nominal overhead costs and no adulteration are the factors might have motivated the respondents towards the PACB. It is appealing to note that a good majority from the entire cross section of the farmers chosen the PACB as their source of input irrespective of their classification. Nonetheless a notable percentage (21.2) of the respondents were not opted the PACB for their input requirement due to the fact that advance money, supply of inputs for credit, non-availability in the PACB, farm gate supply from the non-institutional agencies. Chi square was applied to find the relationship between the purchase of inputs from PACB and classification of the farmer. The calculated chi square value is .742 which is lesser than the table value of 5.991 at 5% level of significance which prove that there is no significant relationship between the two variables.

A considerable size of the respondents (32 percent) opined that less price is the prime motivating reason to prefer PACB for their input requirement followed by easy availability (29 percent) and good quality (26.6 percent). It is encouraging to view that 12.4 percent of the respondents preferred cooperatives for its quality supply of inputs.

The mean score for organic input was 35.40 which is slightly higher than that of inorganic input score (32.83). In fact it is well known factor that the cost of production for organic input is quite higher than the inorganic input likewise the productivity is also equally high in organic input. The T-test was applied to find whether there is significant difference between organic input and inorganic input in the average coconut production cost per acre. Since the calculated value is lesser than the table value it is inferred that there is no significant difference between organic input and inorganic input in the average coconut production cost per acre.

The average coconut produced per acre through single farming is 8.97 which is slightly higher than that of mixed farming (8.56). The T-test was applied to find whether there is significant difference between single farming group and mixed farming group in the average quantity produced per acre. Since the calculated value is less than the table value it is inferred that there is no significant difference between single farming and mixed farming group in the average quantity produced per acre.

The average production per acre is high among medium farmers (9.35) than small (9.32) and big farmers (7.91). One way ANOVA was applied to find whether there is significant difference among the type of farmer groups in the average quantity produced per acre. The ANOVA result shows that there is significant difference among the type of farmer groups in the average quantity produced per acre.

Marketing of Coconuts: Practices and Problems in the Study Area

A very good number of the respondents (272) disposed their coconut with husk. Within this category a sizable portion of the place was occupied by the small and medium category farmers. Value addition is not commonly practiced among the small and medium farmers due to the fact that paucity of manpower, less financial capacity, lack of infrastructural facility, avoidance of time delay in selling, minimizing the chance of price reduction, poor knowledge on technology etc., and rest of the (34) respondents are capitalizing and utilizing the benefits of husk by way of disposing their coconut in the form of without husk which is mostly practiced by the big farmers.

A considerable percentage of the respondents (49.7 %) are falling under the category of opportune farmers which provides ample freedom to follow their own disposal strategy. On the other hand those who are opted for leasing (22.2 %) and debt-bonding (10.5%) don't have a right to decide the disposal strategy. Nevertheless 17.6 percent of the respondents are following mixed practice of disposal for the purpose of capitalizing the market situation. Further scrutiny of the table shows that a sizable portion of the small and medium farmers are under the clutches of leasing and debt –boding system due to the fact that their less financial holding capacity on the contrary a good majority of the big farmers are dictating the market by way of following opportune system.

An attempt has been made to correlate the two variables viz., type of farmers on the basis of mode of disposal and their number of years of experience in coconut cultivation for which the result shows that there is no significant relationship between these two variables.

Most of the respondents leased out their trees due to the financial constraints and other personal reasons like old age and migration to town and other places. Among the reasons for leasing migration occupies a major share. A considerable portion of the respondents (29.4%) leased out their trees for the purpose of meeting out marriage expenses of their children and family members. It is a tradition in India, that the marriages are conducted for the sake of exposing their wealth and to enhance their image among the public. This superfluous practice leads the farmers to end up with more financial burden followed by leasing out coconut trees.

A good majority (65.4%) of the respondents was following wholesale practice due to avoiding Herculean task of selling in small portions, to make lump sum amount of returns, minimize the wastage, lack of manpower to do sorting and selling etc. Nevertheless 14.7 percent of the respondents taken the challenge of direct selling to consumers. It is witnessed from the survey that majority of the small farmers are undergoing direct sale to consumer. This may be due to the fact that small numbers of production, capitalizing the maximum benefit, avoid overhead costs like loading, unloading, transportation etc. The calculated chi square value confirms that there is no significant relationship between the category of farmers and their sale practice.

A simple majority of the respondents (33.1%) has chosen traders as their channel member followed by cooperative marketing society (27.4 %). Flexibility in trade operations, advance payment, farm gate purchase, less cumbersome trade process might have influenced the farmers to go behind the traders. Standardized practice, warehousing facility, advance against the pledge, reasonable price, open and free sale process might have attracted the farmers to have trade relationship with the cooperative marketing society. Nevertheless, rest of the non-institutional agencies like commission agents, wholesalers, industrial user and local marketers also attracted the farmers in a notable way.

The survey results clearly reports that the farmers fetch maximum of Rs.4.00 for both the varieties during flush season. The mean price for grove is slightly higher (3.5521) than periphery (3.3678). Indeed it is a common knowledge that the prices are generally near to the ground during flush season on the contrary the prices are sky-scraping in off season. It is clearly understood from the survey that irrespective of the variety the minimum (Rs.3.00) as well as maximum (Rs.5.00) price is same during the off season.

A good majority of the respondents (52.9 %) were satisfied over the price offered by the respective buyers followed by 35.3 percent of the respondents were not satisfied with the price. It is interesting to note that, among the well educated category a good majority of the respondents (72.7%) were satisfied with the price. This may be due to the fact that these educated respondents may know the market conditions and prevailing market rate for their products.

A good majority of the 196 respondents (64.1 percent) opined that the price offered by the institutional agency is better than that of the non-institutional agencies. This may be due to the fact that institutional agencies like cooperative marketing society and regulated markets are supported and governed by the government and moreover these institutions always in the side of farmers whereas the non-institutional agencies are profit centered and tries to exploit the farmer community as much as possible. Nevertheless about 110 respondents (35.9 percent) have favoured their opinion towards non-institutional agencies.

Though there are various problems with the non-institutional agencies, a notable size of the respondents (35.9 percent) chosen non-institutional agencies for their marketing operations. Further analysis clearly expresses, a simple majority of the respondents

(36.4 percent) opined that advance money is the primary reason for opting non-institutional agencies followed by farm gate sales and no cumbersome process are the reasons felt equally by the 20.9 percent of the respondents. Nonetheless 5.5 percent of the respondents opined that better price was offered by the non-institutional agencies.

A noticeable portion of the (26.8 percent) respondent opined unethical marketing practices is one of the major problems faced by them followed by absence of stabilized price (26.1%) and ineffective institutional agencies (25.5%). This may be due to the fact that the study area is still dominated by the non-institutional agencies and their prime motto is to capitalize the market opportunity not the welfare of the coconut farmers. It is very distrusting factor that institutional agencies like cooperative marketing society and regulated market are also ineffective in the study area.

Coconut Growers' Perception about the Present Marketing Functionaries

The mean score of perception of big farmers (4.52) is higher than that of small and medium farmers towards Cooperative Marketing Society. This may due to the fact that all the big farmers have a strong influence in the operation of marketing society and their capacity to transport the material to the market yard. The Analysis of variance to test the group differences is shown in the table. The F-test is significant at 5 % level which confirms that there is no significant difference among the group towards to cooperative marketing society.

The mean score of perception of medium farmers is slightly higher than that of the small and big farmers. The F- test value is at 5 % level of significance which shows that there is no significant difference among the group in the average preference score towards the regulated market.

The mean score of perception of big farmer (8.95) is quite higher than that of medium (6.03) and small farmer (4.86). Thus one can conclude that there is a strong trade relationship between the big farmers and traders. Flexibility in the trade operations, farm gate sales and transport facility might have influenced the big farmers to have trade relationship with traders in spite of their membership in the cooperative marketing society. The Analysis of Variance to test the group difference is shown in the table. The calculated F -test value is (15.050) confirms that there is significant difference among the groups in the average preference towards the traders. Hence HO_6 : "There is no significant difference among the type of farmer groups in the average preference score towards traders" is rejected.

The mean score of perception of big farmer (9.35) is marginally higher than medium farmer and small farmer. The F- test value is 18.409 which is greater than the table value of 4.676 at 1 % level of significance. Since the calculated value is greater than the table value it is inferred that there is significant difference among the groups in the average preference towards to commission agents.

The mean score of perception of medium farmer is higher (6.39) than the other categories. The calculated F-test value is (8.447) at 1 % level of significance which indicates that there is significant difference among the groups in the average preference towards the wholesalers.

The mean score of perception of small farmers respondents is higher than (6.83) medium and big farmer respondents towards local marketers. Small number of production, no overhead cost, and immediate sale are the factors might have motivated the small farmers to have trade relationship with the local marketers. The ANOVA result shows that the calculated F ratio value is 25.854 which is greater than the table value of 4.676 at 1 % level of significance. Thus it can be concluded that there is no significant difference among the farmer groups in the average preference towards the local marketers.

It is clearly understood that the mean score of perception of small farmer respondents (6.02) is higher than medium (4.18) and big (3.40) farmer respondents. The calculated F ratio value is 6.244 which is greater than the table value of 4.676 at 1 % level of significance. Thus it can be concluded that there is significant difference among the farmer groups in the average classification of the farmer in the industrial user.

The mean score of the quantity and value of coconut marketed by the traders is 40046.4847 and 158508.1081 respectively which occupies first place among the market functionaries. The standard deviation score for traders are 24492.4400 and 105339.7181. Whereas the mean score for the quantity and value of coconut marketed by the cooperatives is 52265 and 182927.5 and the standard deviation is 46632.5812 and 163214.1343.

It is clearly evident from the study that majority of the sample farmers (56.3 %) acquainted with the technical aspects of coconut production from the experienced farmers followed by Government institutions (39.5%). However, the role played by the traders in this regard found to be insignificant.

A simple majority of the respondents (33.1%) chosen traders as their channel member rather than cooperative marketing society (27.4 %). Flexibility in trade operations, advance payment, farm gate purchase, less cumbersome trade process might have influenced the farmers to go behind the traders. Standardized practice, warehousing facility, advance against the pledge, reasonable price, open and free sale process might have attracted the farmers to have trade relationship with the cooperative marketing society. Nevertheless, rest of the non-institutional agencies like commission agents, wholesalers, industrial user and local marketers also attracted the farmers in a notable way.

Majority of the respondents (52.9) have satisfied the price offered by their marketing channels. 60.4, 53.8, 44.3 and 52.9 percent of the respondents have satisfied the price offered by traders, commission agents, industrial users and cooperatives respectively. None of the respondents has satisfied with the local marketers. The calculated value of chi square is 20.352 which are greater than the table value of 20.090 at 1% level of significance. Since the calculated value is greater than the tale value it is inferred that there is significant relationship between the marketing channels and price offered by them. Hence HO_2 : "There is no significant relationship between marketing channels and price offered by them" is rejected.

A considerable majority of the respondents (29.8 percent) opined that the good marketing practices are followed by the traders. Next to the traders Cooperative Marketing Society occupies a place in the hearts of the sample farmers (20.1 percent). Nonetheless Commission agents (19.4 percent) and wholesalers (13.4 percent) are also following good marketing practices in their marketing operations.

A good majority of the respondents opined that the wholesalers do justifiable job (27.8 percent). Rest of the sample respondents favoured their opinion towards reasonable grading to the following market functionary's viz., traders (19.3 percent), commission agents (17 percent), industrial user (15 percent) and cooperative marketing society (8.5 percent).

Majority (60.8 %) of the respondents found Cooperative Marketing Society as the best service oriented marketing functionary. It is distressing to note that Regulated Market being the government run institution didn't make much headway as the response rate among the respondents was found to be insignificant. Nevertheless, 34 percent of the respondents gave due consideration on the service orientation of non-institutional market functionaries.

In spite of different opinion among the farmers about the non-institutional agencies, they like to continue trade with them only as it was witnessed from the table 3 that as high as 84.5 % of the respondents desired to continue trade with them. This might be due to the fact that non-institutional functionaries as compared to institutional functionaries seemed to adopt high flexibility in terms of rule, procedures and regulations.

Though the non-institutional agencies hold giant share in the volume of business with the coconut farmers, they are in the problematic side. Most of the respondents (32.4 percent) felt that commission agents are the most problematic in the market. Only 9.8 percent of the sample respondents inferred that cooperative marketing society is also one of the problematic market functionaries in the coconut market.

The survey result revealed that majority of the respondents opined that timely payment is being effected by traders, local marketers, industrial user and commission agents.

As to credit requirements, the survey result showed that majority of the short- term financial needs of farmers is being met out by the non-institutional agencies (92.8 percent).

It is observed that majority of the farmers (30.3 percent) have obtained long-term loans from the wholesalers. Industrial user (17 percent), local marketers (16 percent), traders (15.7 percent) also played a considerable role in issuing long-term loans to the coconut growers. (Table 6.20)

The survey result shows that non-institutional agencies (94.3%) have been playing a significant role in pre-procurement process.

It was observed that 25.5 percent of the respondents have high opinion towards traders on post procurement practices which may be due to the facts like supply of fertilizer at the door steps, provision of finance in advance, personal intimacy etc. On the other hand local marketers also play a considerable role (24.2) in providing post sales service to the coconut growers.

It is clearly understood that 34.7 percent of the respondents opined that commission agents do various malpractices during their marketing operations followed by 20.2 percent traders. Malpractices in Cooperative Marketing Society are found to be low as compared to other marketing functionaries.

It is observed that majority of the respondents (23.8 percent) opined that commission agents are the only nearby source to market their produces. A very few proportion of the respondents viewed their opinion towards local marketers and wholesalers.

It is evident that most of the farmers (24.1 percent) felt that trader follows hassle free trade and followed by wholesalers (20.6 percent). Rest of the respondents felt that commission agents (18.6 percent), industrial user (15 percent) and local marketers (14.3 percent) follow less cumbersome process in their marketing activities. Only 3.8 percent of the sample respondents opined that cooperatives also follow hassle free trade in their marketing operations.

It is clearly understood that majority of the respondents (24.5 percent) felt that cooperative marketing society provides godown facilities for their produces. Wholesalers (19.3 percent), local marketers (17 percent) and industrial user (12.4 percent) are also kind enough to extend their space for the storage purpose of the farmers.

Most of the farmers (22.5 percent) felt that wholesalers provide enough market information about the market. Next to the wholesalers traders stood second (21.7 percent) in providing relevant market information to the farmers. Local marketers (19.5 percent), commission agents (13.4 percent) and industrial user (12.7 percent) also plays a considerable role in providing market information. Only 8.3 percent of the respondents said that cooperative marketing societies also provide some market information to the respondents.

It is evidenced that among the non-institutional agencies commission agents have high price fluctuation in their marketing operations (33.4 percent). Next to commission agents, the respondents felt that there are high price fluctuations among the wholesalers (17 percent) and local marketers (17 percent). Only 6.5 percent of the respondents felt that there is a high price fluctuation in cooperative marketing society.

Majority of the respondents (31 percent) opined that commission agents lack some standardized practice in their marketing operations. Next to commission agents, traders stood second (21.3 percent) in lacking standardized practice. Rest of the sample population opined that wholesalers (16.7 percent), local marketers (12.7 percent) and industrial user (8.2 percent) lacks in practicing standards in their marketing operations. Only 8.8 percent of the respondents opined that cooperative marketing society also lacks standardized practice in their marketing operations.

Most of the respondents (21.3 percent) opined that value addition services are more in the industrial user. Commission agents also took part considerably in value addition activities (19 percent). A good majority of the respondents (17.5 percent) gave their consent towards traders in this value addition process. Out of 306 samples only 30 sample populations (9.9 percent) viewed that cooperative marketing society also undertakes value addition activity.

It is cleared that majority of the respondents (39.3 percent) said that the cost of sale is high in cooperative marketing society. This is may be because of the institutional agencies like cooperatives and regulated market cannot adopt the practices followed by the non-institutional agencies like free transportation, farm gate pick up etc. 17.3 percent of the respondents opined that there is a high cost of sale they face while marketing their produces to the wholesalers. Rest of the respondents opined that there is a high cost of sale among the commission agents (14.4 percent) and traders (14 percent).

It is inferred that majority of the respondents (27.1 percent) availed pledge loans from the cooperative marketing society. It is observed that commission agents played a considerable role (19.9 percent) in issuing pledge loan to the coconut peasant community. It is notable that the Wholesalers (16.4 percent) and traders (12.4 percent) also have a practice of issuing pledge loans to the farmers.

It is clearly understood that most of the sample respondents accepted that the non-institutional agencies (93.1 percent) offered the farm gate purchase. Only 6.9 percent of the respondents said that institutional agencies also do this type of arrangements to the coconut farmers.

It is inferred that a considerable majority of the respondents (23.3 percent) opined that the traders are the most advantageous market functionary in the market. Next to the trader, it is said that commission agents (21.2 percent), local marketers (18.5 percent), industrial user (15.3 percent) and wholesalers (12.3 percent) are also in a position to give some benefit to the farmers. Only 5.8 percent of the respondents opined that cooperative marketing society is one of the most advantageous market functionaries in the market.

From the analysis it is known that majority of the respondents (23.7 percent) opined that the traders have high technical knowledge among the market functionaries. 22.2 percent of the samples have an opinion that commission agents also possess technical know how about the coconut market. Only 3.6 percent of the respondents favoured towards cooperative marketing society in this regard.

A good majority of the sample respondents (26.7 percent) opined that traders follow somewhat open and free sale proceeds in their marketing operation. 21 percent of the respondents said that commission agents also follow this type of open and free marketing operations. Only 5.5 percent of the sample respondents favoured cooperative marketing society in this regard.

It is clear from the above table that a good majority of the sample respondents (44.5 percent) opined that the overall service is good among the traders. Next to traders, a good majority of the respondents (18.3 percent) inferred that cooperative marketing society is also performing good service to the peasant community. 14.7 percent of the respondents favoured their opinion in this regard to the commission agents. Rest of the 8.5, 7.5 and 6.5 percent of the respondents exercised their right towards industrial user, wholesalers and local marketers.

Conclusion

The coconut industry is growing in terms of production. However, its share in oil and fats trade has consistently declined in the past four decades. Vast growth opportunities remain for the coconut industry, but the marketing strategy needs a reorientation to suit to emerging trend. Awakening and alertness on diversification of coconut, with a motive to recapture the market have to be provided priority. The best option with vast growth opportunities for product diversification and value addition exists. Despite this vast potential, the industry can flourish only through strategic initiatives and synergy among the organised and unorganised market outlets function. Strategic marketing has to include product diversification, marketing intelligence backed by market research to widen the market base in different coconut growing region. Cooperation from the organisations and upfront approach is needed in partnership mode to address the issues in penetrating the existing markets.

Since the existing marketing networks are dominated and controlled by private marketing functionaries, farmers are deprived of the major chunk of the consumers' price for coconuts. Farmers know each and every disadvantage of sale through the existing channel members. However, they depend on them by compulsion. Non-availability of organised credit for production, marketing and consumption purposes, absence of organised marketing institutions and difficulties in the adoption of off-farm value addition, processing and marketing technologies on coconuts are the major reasons for the 'forced sale' of coconut either through lease practices or on-farm sale of coconuts to marketing intermediaries. They, therefore, demand rejuvenation of organised marketing institutions such as cooperatives and regulated markets and promotion of a network of micro-enterprises of coconut farmers facilitating for collective bargaining and direct marketing of coconuts in places where the organised marketing institutions are defunct.

Suggestions

Based on the findings of the study, it is suggested that the following measures may be taken by the Government of Tamil Nadu to safeguard the interest of the coconut growers in the sample area.

- Since the study area is coconut concentrated, formation of exclusive coconut producers' cooperative marketing society in the lines of turmeric producers' cooperative marketing, groundnut producers' cooperative marketing society and Sago producers' cooperative marketing society in and around the Pollachi may solve the marketing problems of coconut growers.
- The coconut growers in the sample area are lacking in availing the institutional credit for their regular maintenance of coconut gardens. Some sort of organized finance is the need of the hour. If not the sample coconut farmers always under the clutches of non-institutional agencies.
- Heavy price fluctuation causes unexpected loss in the expected income. To stabilize the price of the coconut and its products the Government should come forward for the implementation of price guarantee scheme.
- Absence of coconut based processing industries in the study area is one of the reasons for the heavy price fluctuation in the coconut industry. Hence, Government of Tamil Nadu should come and take necessary steps to promote coconut based cottage and village industries and small scale units.
- Among the two institutional agencies in the study area, only one institution namely Pollachi Cooperative Marketing Society Ltd commences with copra

procurement in collaboration with the National Cooperative Marketing Federation Ltd (NAFED) that too for a very limited period of time. The Anamalai Regulated Market is another institutional agency in the study area but this institution didn't deal with coconut marketing. Hence, the Government of Tamil Nadu must take essential steps to revitalize the institutional marketing agencies for the betterment of coconut farmers in the study area.

- Since the study area is intense coconut area both the Tamil Nadu and Indian Government may take successful steps for the establishment of integrated coconut processing complexes which will definitely augment the standard of living of coconut growers.

Bibliography

1) Adhikari, S., Dwivedi, S., Luthra, S. and Narayana, D.B.A. (1996). Coconut Oil as a Hair Oil, In Seminar Papers, National Meet on Coconut Oil, Coconut Development Board, Kochi.

2) Ajith Kumar, P., (1994). Coir Industry in India (Processing and Marketing of Coconuts in India), CDB, Kochi.

3) Anirudha Sharma (1993), Villagers can do well with better Marketing, *Indian Farmer Times*, Vol. 10, No. 11.

4) Anonymous. (1989). Report on the meeting of the working group on Coconut-Based Farming Systems, held at Chumpon, Thailand, September 1989. UNDP/FAO project RAS/80/032. FAO (Italy).

5) __________. (2001). Report on Sample Survey for Estimation of Area and Yield of Coconut and Arecanut in Tamil Nadu, Dept. of Economics and Statistics, Chennai.

6) Aravindakshan, M. (1996) Coconut Development in 2000 AD, Coconut Development Board, Kochi.

7) __________. (1996). Challenges to the Coconut Industry in India and Strategy for making it competitive, Coconut Development Board, Kochi.

8) __________. and Nandakumar, T.B. (1996). Coconut in Social and Cultural Life of People in India (in *Coconut for Prosperity*, Eds. P.K.Thampan).

9) _________, (1997), India attains the premier position in coconut production, *Indian Coconut Journal*, Kochi; CDB, Vol. 28, No.3

10) _________, (1995). Emerging Trends in Coconut Culture and Challenges to Research, *Indian Coconut Journal*, CDB, Kochi, (Vol.26, No.9).

11) __________. (1997). India Attains the Premier Position in Coconut Production, *Indian Coconut Journal*, CDB, Kochi, (Vol.28, No.3).
12) Arboleda, No. P., Floria, R.N., and Candy, C.S. (1986). Coconut-problems and approaches. The Philippine experience. Paper for workshop proceedings of the international consultation on Plantation Crop Diversification. FAO (Italy), November 1986.
13) Bavappa K.V.A. (1976). CPCRI and Coconut Research, *Indian Farming*, 26, (9).
14) __________, Kailasam, C., Khader, K.B.A., Biddappa, C.C., Khan, H.H., Kasturi Bai, K.V., Ramadasan, A., Sundararaju, P., Bopaia, B.M., George V.Thomas, Misra, L.P., Balasimha, D., Bhat, N.T. and Sham Bhat, K. (1986). Coconut and Arecanut based Cropping Systems, *J.Plantn. Crops*, 14(2).
15) Bernd Kretschemer and Amal Kumar Bose, (1999). Protein shifting in Coconut processing. *Indian Coconut Journal*, 30, 36-37p.
16) Bhatia, B.M. and others (1974), "Food, Famine and Population", Population in India's Development.
17) Boceta Norberto, M. (2000). The Challenges of making the coconut industry globally competitive. In Souvenir, International Coconut Conference 2000. Coconut Development Board, Kochin.
18) CDB, (1997), *Coconut Situation in India*.
19) __________. (1994). Proceedings of the National Seminar Report on Processing and Marketing of Coconuts (SPAMCO) held on 13th January, Bangalore.
20) __________. (1999). Integrated disease management in coconut, *Indian Coconut Journal*, Vol. XXIX, No. 4, Aug.
21) __________. (2001), Tender Coconut, Kochi.
22) CFTRI, 1999. Home Scale Processing and Preservation of Fruits and Vegetables. CFTRI, Mysore, 11-12p.
23) Chandrashekhar, G. (2003), Edible Oil adulteration-lax food law to blamc, Bombay: *Business Line*, June, 9.
24) Chettiar Lakshmana, S.M.L. (1973). *Folklore of Tamil Nadu*, National Book Trust, New Delhi.
25) Child, R. (1953). *New Biology*, The Coconut, Longman Group Ltd., London.
26) Coconut Development Board. (2001). *Tender Coconut*, Coconut Development Board, Kochin.
27) __________, (2003). Demand and supply of coconut products and consumers buying behaviour in Tamil Nadu; Research Report.
28) Coir Board. (2000). Annual Report, Kochi.
29) CPCRI. (2000). Annual Report, Central Plantation Crops Research Institute, Kasaragode, Kerala.

30) Dwarakanath. (1994). Vinegar Fermentation with Special Emphasis on Possibilities of Utilisation of Matured Coconut Water (Processing and Marketing of Coconuts in India), CDB, Kochi (pp.44-51).

31) Edachal Mohammed. (1963). Coconut Stories, Coconut Bulletin, 17 (6).

32) FAO, (1960), Marketing of Agricultural Products, Rome, p. 16.

33) George, M.V., Vijaya Kumar and others (1991). Trend in Area, Production and Productivity of coconut in India, *Indian Coconut Journal*, Kochi, Vol. XXII, No.647.

34) Gnanasekar Francis. (2003). Globalisation and WTO – Indian, Insight, *Kisan World*, March.

35) Haridoss and Chandran (1996). Marketing systems, costs, margins, price spread and marketing problems of coconut- a case study of coconut growers and traders in Tamil Nadu, *Indian Coconut Journal*, Vol XXVII, Nov.1, May

36) Hough, E.M. (1966). *The Cooperative Movement in India*, Calcutta: Oxford University Press

37) Jnanadevan, R. and Prakash, R., (1993). An analysis of constraints in increasing coconut production in Kerala, *Indian Coconut Journal*, No. 4, August.

38) Jon J. Kabara. (2000). Health oils from the tree of life-nutritional and health aspects of coconut oil, Indian Coconut Jouanal, No.8, December.

39) Kadalli and Suseela Nair. (2000). Manurial value and efficiency of coir dust based enriched super compost, *Indian Coconut Journal*, Vol.XXXI, No.3, July.

40) Kasembe, J.N.R. and Currah, I.A. (1968). Coconut experiments in Chambezi, Tanzania, Paper, FAO Technical working party on Coconut Production, Protection and Processing, Jakarta, Indonesia.

41) Katar Singh. (2002). Cooperatives – Challenges in emerging economy, NCDC Bulletin, December.

42) Kaul, G.L. (1994). Product Diversification in Coconut – World Scenario (Processing and Marketing of Coconuts in India), CDB, Kochi.

43) Kaushal, H. (1994). Optimum Utilisation of Coconut (Processing and Marketing of Coconuts in India), CDB, Kochi.

44) Lalithamma, K. (1996). *Pharmacopoeia*. Government Ayurvedic College, Thiruvananthapuram.

45) Lebai Mat, H. (1979). Coconut Smallholders development scheme. *Community Quartely Supplement*.

46) Liyanage, D.V., Mankey, T., Luntungan, M., Djisbar, A., and

Sufiani, S. (1986). Coconut Breeding in Indonesia. Cocos (Indonesia), 4:.

47) Madhavan, K., Bosco, S.J.D. and Sunil Gulati. (1991). Recent Developments in the Harvest and Post-harvest Technology of Coconut (in *Coconut Breeding and Management*, Eds.Silas et. al.), KAU, Trissur.

48) Mall, P.P. (1994). Desiccated Coconut Industry in India – Problems in Internal and Export Trade *(Processing and Marketing of Coconuts in India)*, CDB, Kochi.

49) Marapandian, P. (1994). Role of Kerafed in Coconut Processing and Marketing *(Processing and Marketing of Coconuts in India)*, CDB, Kochi.

50) Markose, V.T. (1992). *Domestic Marketing of Coconut Products in India*, APCC, Jakarta, Indonesia.

51) __________. (1994). Coconut Development in the last decade, *Indian Coconut Journal*, Kochi: CDB, Vol. 25, No. 7, p. 12.

52) __________. (1994). Coconut Processing Sector in India – An Overall Perspective *(Processing and Marketing of Coconuts in India)*, CDB, Kochi.

53) __________. (1995). *Coconut Production and Productivity in India* (APCC Publication).

54) __________. (1998). Fifty years of Coconut Development in India (in *Indian Coconut J*.29).

55) Martin et. al., (1979). cited in *Coconut, Tree of Life*, Ed: Ohler, J.G., 1984., FAO, United Nations.

56) Mary, G. (1999). Rationale for use of lauric oils as physically functional foods for HIV infected individuals, *Indian Coconut Journal*, September.

57) Mathew, A.C. et. al., (2003). Energy Efficient Irrigation System Design, *Kisan World*, June.

58) Memoria, C.B., (1998). *Agricultural Problems of India*, Kitab Mahal, Allahabad.

59) Menon, K.P.V. and Pandalai, K.M. (1958). The Coconut Palm – A Monograph, Ernakulam: Indian Central Coconut Committee.

60) Muralidharan, K. and Remani Gopalakrishnan, P. (1998). Coconut Development in India – The Past, Present and Future (in *Indian Coconut J*. 28).

61) Nair, M.K., Rathinambal, M.J. and Gopalasundaram, P. (1996). Coconut Research in India (in *Coconut for Prosperity* – Eds. Thampan, P.K.),

62) Namboodhiri, K.U.K. and Singh, H.P. (2000). *Trends in Coconut Research and Development in India*, CDB, Kochi.

63) __________. and Nair, G.S. (1998). Nattarivukal. *Karshakashree,* August, 1998. p.34-35.

64) __________. (1998). Milestones in Coconut Research, *Indian Coconut Journal,* May.

65) Nandakumar, T.B. (2000). *Pleasant Fluid of Coconut in Souvenir,* International Coconut Conference, Coconut Development Board, Kochi, India.

66) __________. (1985). Coconut palm products in social and religious functions in India, *Coconuts Today,* 3 (2).

67) __________. (1997). Thenginkallu. *Indian Nalikera Journal,* April, 1996. p.10-11.

68) Norberto M. Boceta. (2000). The Challenges of Making the Coconut Industry Globally Competitive (in Souvenir, XXXVII COCOTECH MEETING, CDB, Kochi).

69) Ohler John. G. (1999). The Coconut Palm and its Environment–Historical Background (in *Modern Coconut Management – Palm Cultivation and Products*), FAO of UN, New Delhi).

70) __________. (1984). Coconut, Tree of Life, FAO of the UN. Rome.

71) Paget, R.A.S. (1929). Origin and dispersal of palm in America, Cont. U.S. Nat. herbs. XIV.

72) Panicker, T.R.R.K. (1960). Brihath Asavarishta Manjula. Shree Lakshmivilasam Book Depot, Ernakularm.

73) Patel, J.S. (1938). The Coconut – A Monograph. Madras: Supt. Press, Madras.

74) __________. (1938). The Coconut- A Monograph. Madras: Supt. Press, Madras.

75) Pillai, Gopala, A.S. (1982). *Ayurveda Oushadha Gunachandrika,* Quilon: S.T.Reddiar and Sons, V.V. Press, Quilon.

76) __________. (1979). Pazhangalum Oushodagungaium, (Mal) 2nd ed., Quilon: S.T.Reddiar and Sons, V.V. Press, Quilon.

77) Poduval Sreekumar. (2002). Diversification and Value addition in Coconut Processing (in *Sustainable Production and Utilisation of Coconut*) (Eds. H.P Singh and M.T.Mathew), CDB, Kochi.

78) Punchihewa P.G. (2000). Current Status of the Coconut Industry, Chennai: 37th Cocotech Meeting/International Coconut Conference, July.

79) Rama Varier Kaikulangara. (1961). *Arogyakalpadruma Balachikitsa,* Ed. Pub.: S.T. Reddiar and Sons, Quilon.

80) Ramakrishna Pillai, K. (1994). Role of Co-operatives in Processing and Marketing of Coconuts in Kerala with Special Reference to MARKETFED (Processing and Marketing of Coconuts in India), CDB, Kochi.

81) Rathinam, R. and others, (2002). Global scenario of coconut oil, Paper presented at OFIC Kula Lumpur, Malaysia, October.

82) Ratnambal, M.J., Arunachalam and Krishnan. (2003). Floral Biology of some coconut accessions. *Journal of Plantation Crops*. 31 (1).

83) ________. (1999). Varieties suitable for Tender Coconut. *Indian Coconut Journal*. 30, 64-66p.

84) Rethinam, P. and Nandakumar, T.B. (2001). *Coconut Statistics* (Supplement), CDB, Kochi.

85) Rethinam, P. (1991). Management of Coconut Palms (in *Coconut Research and Development*, Eds. Silas et.al.,), KAU, Vellanikkara.

86) ________. (2001). *Coconut Industry in India*, CDB, Kochi.

87) ________. (2001-a). *Operational Guidelines* - CDB Schemes - IX Five Year Plan CDB, Kochi.

88) Ron Harris, (1992). The Coconut Industry- Utilization and Marketing, *The Planter*, Kuala Lumpur, Malaysia.

89) Satyasunaram, I. (1997). *Market survey-Coconut Palm: Diversification holds the key, Facts for you*, New Delhi: Indian Book House, Vol. 18, No.5, September.

90) Satyavati Krishnankutty. (1994). Products from Coconut Water (*Processing and Marketing of Coconuts in India*), CDB, Kochi.

91) Sharma, Amod. (2001), Muscling up agricultural markets, *Agriculture Today*.

92) Shivanandiah, T.M. (1970). Use of tender coconut water in gastro enteritis, *Cey. Coco.J.*, 21.

93) Singh Sohinder Bedi. (1971). Folklore of Punjab, National Book Trust, New Delhi.

94) Singh, H.P. (1998). Coconut Industry in India – Challenges and Opportunities, CDB, Kochi.

95) ________. (2002). *Operational Guidelines*, TMOC, CDB, Kochi.

96) ________, (2002). Strategic Agenda for Improving Comptitiveness of Coconut Industry in the Midst of Challenges from Competing Products (paper presented in XXXIX Cocotech, Thailand).

97) ________, and Mathew, M.T. (2002). Sustainable Production and Utilisation of Coconut, CDB, Kochi.

98) Sreekumar Poduval, Ajay. M. Pillai and Balachandran Nair, (1998). In : Harvest and Post Harvest Technology of Plantation Crops. Technologies for product diversification and by product utilization in coconut, p. 70.

99) Sreekumar Poduval, (1994). Utilisation of Coconut by-Products in India (*Processing and Marketing of Coconuts in India*), CDB, Kochi.

100) Srivastava, J.N.L. (2002). Strategic efforts taken in India to make Coconut Industry Globally Competitive (paper presented in XXXIX Cocotech, Thailand).

101) Subburaj, B. (1996), Coconut Processing industry in Tamil Nadu, with special reference to Kangayam, Paper presented in the National Meet on Coconut, held at Kochi, November.

102) ________. (1996). Coconut Processing industry in Tamil Nadu, with special reference to Kangayam, paper presented in the National meet on Coconut, held at Kochi, November.

103) ________. (1997). Coconut Oil Mills in Kangayam Region, Erode District, Tamil Nadu, *Indian Coconut Journal*, 28 (7).

104)________. (1998). Participatory Appraisal on Marketing of Coconuts – Practices, Problems and Prospects in Tamil Nadu, GRI, Gandhigram.

105) ________. (2000). Coconut Marketing in Tamil Nadu, Ministry of Rural Areas and Employment, Department of Rural Development, GOI, New Delhi.

106) ________. (2004). Coconut Production, Practices and Problems in Tamil Nadu, A Research Monograph, Gandhigram Rural Institute, Gandhigram, Dindigul.

107) ________. (2004). Coconuts in Regulated Markets. Research Monograph, Gandhigram Rural Institute, Gandhigram.

108) ________. (2000). Marketing of Coconuts – Practices and Problems in Tamil Nadu- A Research Monograph. Gandhigram Rural Institute, dindigul, May.

109) Sukhatme, P.V. (1965). *Feeding India's Growing Millions,* quoted in Memoria C.B.

110) Thampan, P.K. (1996). Coconut for prosperity. Peekay Tree Crops Development Foundation, Kochi, Kerala, 247p.

111) ________. (1975). *Handbook on Coconut Palms,* Oxford and IBH Publishing Co. Pvt. Ltd., New Delhi.

112) ________. (1988). *Glimpses of Coconut Industry in India,* CDB, Kochi.

113) ________. (1993). *Hand Book on Coconut Palm* (3rd edition), Oxford and IBH Publishing Company, New Delhi.

114) ________. (1994). *Facts and Fallacies about Coconut Oil,* APCC, Jakarta.

115) ________. (1996). *Coconut for Prosperity,* Peekay Tree Crops Foundation, Kochi.

116) ________. (1998). The Coconut Profile of India (in *Glimpses of Coconut Industry in India*), CDB, Kochi.

117) ________. (1988). *Glimpses of Coconut Industry in India,* CDB, Kochi.

118) __________. (1995). *Global Competitiveness of Coconut Industry*, Kochi: Pekay Treecrops, Kochi.

119) Thomas, P.T. (2002). Marketing prospects of coconut products in A and N Islands, Paper presented in the High power Committee held at Port Blair, May.

120) Vaidyan Padmanabhan, N.K. (1996). cited in *Coconut for Prosperity*, Ed. Thampan, P.K., Peekay Tree Crops Foundation, Kochi.

121) Venkatachalam V. (1997). Investment opportunities in coconut in Tamil Nadu, Paper presented in the CII, Madurai. Indian coconut Journal, June. p.8.

122) __________. and Thomas, P.T. (1997). Investment opportunities in Coconut in Tamil Nadu, paper presented in the CII, Madurai. *Indian Coconut Journal*, June.

123) Warrior, P.K. (1996). *Coconut in Ayurveda*, Peekay Tree Crops Development Foundation, Kochi.

Appendices

APPENDIX I

COCONUT AREA IN COIMBATORE DISTRICT

(TALUK WISE - 2005-2006)

S.No.	*Name of the Taluk*	*Area in Hectares*
1.	Pollachi	49660
2.	Udumalpet	23518
3.	Palladam	10782
4.	Coimbatore (South)	7597
5.	Tirupur	4027
6.	Coimbatore (North)	2757
7.	Avinashi	1659
8.	Mettupalayam	1541
9.	Valparai	Nil

Source: District Statistical Hand Book, Year: 2005-2006.

APPENDIX II

COCONUT AREA IN COIMBATORE DISTRICT

(BLOCK WISE - 2005-2005)

S.No.	Name of the Taluk	Area in Hectares
1.	Anamalai	16999.435
2.	Pollachi (North)	13357.455
3.	Udumalpet	11331.390
4.	Kinathukadavu	10205.605
5.	Gudimangalam	9318.005
6.	Pollachi (South)	9097.535
7.	Sulthanpet	7085.815
8.	Madukarai	4533.965
9.	Pongalur	3411.675
10.	Thondamuthur	3063.400
11.	Madathukulam	2868.815
12.	Sulur	1990.515
13.	Palladam	1705.820
14.	Periyanaickenpalayam	1650.020
15.	Karamadai	1540.515
16.	Sarkar Samakulam	1106.870
17.	Avinashi	876.655
18.	Annur	697.000
19.	Tiruppur	615.090

Source: District Statistical Hand Book, Year: 2005-2006

APPENDIX III

COCONUT AREA IN ANAMALI BLOCK (VILLAGE WISE - 2005-2005)

S.No.	*Name of the Taluk*	*Area in Hectares*
1.	Vettaikaran Pudur	2975.415
2.	Marichinaicken Palayam	2069.565
3.	Peria Podu	1771.930
4.	Kottur	1762.520
5.	Kalia Puram	1657.685
6.	Odayankulam	1499.885
7.	Arthanari Palaym	683.430
8.	Pethanaickanur	663.370
9.	Kambaiapatti	628.195
10.	Jalli Patti	575.355
11.	Anamalai	554.835
12.	Thuraiyur	363.405
13.	Angal kurichi	362.715
14.	Somathurai	350.800
15.	Pilchinnampalayam	255.895
16.	Athupollachi	247.845
17.	Karianchetti Palayam	245.175
18.	Thensangampalayam	203.590
19.	Thenchittur	127.825

Source: District Statistical Hand Book, Year: 2005-2006

APPENDIX IV

COCONUT DEVELOPMENT BOARD

Directorate of Coconut Development

This period was followed by a Plan break of three consecutive financial years viz. 1966-67, 1967-68, 1968-69. The Directorate mainly undertook planning and co-ordination of central and state sector programmes on coconut development in the country. The Directorate functioned during full term of the Fourth and Fifth Plan periods and the first year of the Sixth Plan, which commenced in 1980-'81. During these periods short and long term production and productivity improvement programmes were continued with major emphasis on hybrid planting material production and rejuvenation programme of the disease affected coconut holdings in view of the root (wilt) disease. For assisting in the planning process, an advisory committee viz. Indian Coconut Development Council was also constituted giving representation to varied interests related to the industry.

Development programmes of coconut received attention under the Five Year Plan programmes of the country. Different short term and long term production programmes were envisaged and implemented with major objective to increase production during this period. Laying out demonstration plots, expansion of irrigation facilities and plant protection were the short term measures implemented. Under the long term programmes, production and distribution of hybrid planting material in the major growing states of the country and subsidized supply of quality planting material were envisaged. Expansions of area under coconut and nursery programmes were also taken up as long term measures.

In the late 1970's, which coincided with the Fifth Five Year Plan of the country, productivity oriented programmes like package scheme, laying out demonstration plots, production and distribution of T x D and D x T hybrid combinations and maintenance of elite farms were introduced in addition to the surveys on cost of cultivation, coconut production and processing aspects. Plant protection measures were given adequate thrust by adopting comprehensive spraying and multiplication of parasites. The Directorate functioned till the formation of the Coconut Development Board in 1981 (Markose, 1998).

Coconut Development Board

The Coconut Development Board is statutory body established by the GOI under an act of parliament in 1979. The Board came into existence on 12th January 1981 and its development programmes commenced in 1982-'83, the third year of the Sixth Five Year Plan. The Board functions under the administrative control of the Ministry of Agriculture, Government of India with its head-quarters at Kochi at Kerala State and Regional officers at Bangalore in Karnataka, Chennai in Tamil Nadu and Patna in Bihar. Therefore 5 state centres located at Agontala, Calcutta, Buvaneswar, Hyderabad and Portblair, besides Demonstration-cum Seed Production farms at Neriyamangalam in Kerala, Vegivada in Andhra Pradesh, Mandya in Karnataka, Abhayapuri in Assam Singheswar in Bihar, Pitappally in Orissa, Kondagaon in Chattisgarh and Belbari in Tripura. The major functions of the Board include adopting measures for the development of coconut industry, recommending measures for improving the marketing of coconut and its products, imparting technical advice to any person who is engaged in the cultivation of coconut or processing and marketing of coconut and its products, providing financial or other assistance for the development of high yielding coconut hybrids, adoption of improved methods of cultivation of coconut, modern technology for processing of coconut and extension of area under coconut cultivation, adopting such measures for assisting coconut growers to get incentive prices, recommending measures for regulating import and export of coconut, collecting statistics relating to coconut industry and publishing the statistics, fixing grades, specifications and standards for coconut and its products, financing suitable schemes so as to increase the production of coconut and to improve its quality and yield, assisting, encouraging, promoting or financing agricultural technologies, industrial or economic research on coconut and its products, undertaking such publicity and publishing such periodicals, books or bulletins on the research and development of coconut and its products, setting up of regional offices and other agencies for the promotion and development of production and grading and marketing of coconut and its products (CDB, 2001).

The development programmes on coconut were given new dimension by the Board by identifying key areas where efforts were to be concentrated. Programmes were formulated with the objective of creating permanent production potential, stepping up productivity, developing appropriate technologies for product diversification and improved marketing of the crop. Expansion of area under coconut, production and distribution of quality planting material, technology development, surveys, extension and publicity activities were the major programmes implemented during the period.

A beginning was made on the development of technologies for product diversification and by-product utilization by the Board and various products developed. While the importance of scientific management practices in coconut was felt essential, establishment of demonstration cum seed production farms in different locations both traditional as well as non traditional belts was taken up as a major activity in the development programmes by the Board.

The root (wilt) disease problem in coconut in the state of Kerala, the prime producer of coconut, aggravated during this period. Eradication of root(wilt) disease advanced palms and replanting with quality seedlings was one of the major recommendations of the research to the farming community. This led the Board to evolve programmes for cutting and removal of disease affected palms by giving compensation to the farmers for cutting and removal and adoption of other improved management practices.

The Board expanded its extension network, through establishing field offices in various locations in the country as well as demonstration cum seed production farms in various states. Concerted efforts of the Board could bring about notable improvement in increasing the overall area, production and productivity level in the country. The Board was also instrumental in introducing the crop in non traditional areas in the North and North-Eastern states of the country.

The Information network of the Board was expanded with the initiation of a computerized coconut information centre at the head quarters with the main objective of exchanging up-to-date information on all aspects of coconut industry among the major coconut growing countries in the world under the Integrated Coconut Information Service Programme initiated by the Asian Pacific Coconut Community. A webpage for the Board has been created in the Internet.

The CDB actively participated in transfer of technology through various mass media. The publications of the Board have played a role for the dissemination of information on all aspects of coconut cultivation and industry. The information transmitted through the Journals on the latest developments in production, protection, processing and marketing of coconut has created awareness among the farming community on scientific cultivation and processing. The Journals also carry monthly market reviews covering the arrivals, price behavior and demand trend for coconut and its product in all the important markets in the country. The publicity and extension activities undertaken by the Board especially in wiping out the misconception on coconut oil among the urban populace and creation of general awareness on coconut products are commendable achievements.

Publicity and extension activities form an integral part of the development programmes of the CDB. Such efforts are needed to educate the farmers, processors, traders and new entrepreneurs. Publications and various other mass media dissemination of information is being carried out by the Board. Documentary films are also produced and screened with a view to educate and inform the farmers, processors and potential entrepreneurs. Production of publications in different languages, production and screening of films, participation in exhibitions and fairs, conducting seminars and workshops, market promotion campaigns, imparting training to unemployed youths in palm climbing and plant protection, training and study tour of farmers to educate them in scientific coconut cultivation are the different extension oriented programmes of the Board.

Apart from the publicity and publication activities, a National Information Centre has been set up with the objective of collection, compilation and maintenance of up-to-date information on all aspects of coconut industry encompassing cultivation, processing, marketing, technology and research at national and international level so that access to and exchange of valuable information is possible within and between the countries (Rethinam, 2001, Rethinam, 2001-a, Singh 2002).

Schemes of Coconut Development Board

Production & Distribution of Planting Material: The objective of the programme is to enhance the production and supply of good quality planting materials through following component programmes.

Establishment of 7 Demonstration cum Seed Production (DSP) Farms in different parts of the country in a total area of 240 ha with the aim to produce quality seednuts. These farms are the demonstration centres for scientific coconut cultivation and processing.

Setting up of coconut nurseries attached to the DSP farms to produce quality seedlings of desired cultivars/varieties suitable for each locality. Raised seedlings are distributed to farmers at reasonable price.

Aid to registered/ private/ approved coconut nurseries. The financial assistance is limited to 25% of the cost of production or Rs.2 lakhs whichever is less. The minimum financial assistance of Rs.50000 is allotted for producing 6250 seedlings annually from 25 cents and maximum financial assistance of Rs.2 lakhs for producing 25000 seedlings from one acre.

Establishment of Regional coconut nursery for supplementing the nursery programmes of the state government. The maximum financial assistance for setting up of regional coconut nursery in non traditional areas is 50% of cost of production.

Aid to establish nuclear seed garden in private sector/registered/ cooperative societies/farmers' associations. Maximum financial assistance is limited to Rs.6 lakhs to establish seed garden phased over a period of 3 years @ Rs.3 lakhs during the 1st year and Rs.1.50 lakhs each in second and third year respectively. The quantum of financial assistance is based on the total area of seed garden (maximum 4 ha) or limited to 25% of the cost of establishment of nuclear seed garden.

Expansion of Area under Coconut

This programme is intended to bring in more area under coconut in the potential belts thereby increasing the production potential in the country. For this, the small and marginal farmers are given incentive assistance for undertaking new planting of coconut and its further maintenance. The new planting assistance is given at the rate of Rs. 8,000 per hectare, which is disbursed in two equal annual installments.

The programme is implemented by the Board through the Department of Agriculture/Horticulture of various States/ UT's as the cases may be. Application forms prescribed by the Board are supplied to various implementing offices of the State Government which in turn are distributed to the eligible farmers. The filled in applications submitted by farmers after proper field verification are forwarded to the office of the Board concerned with proper recommendation of the implementing officers for further processing and sanction of subsidy. The Board after scrutiny of the applications, sanctions eligible subsidy and disburse the sum in the form of individual cheques/ demand drafts to the farmers through the same implementing offices.

Integrated Farming for Productivity Improvement

The objective of the programme is to improve production and productivity of the coconut holdings through an integrated approach and thereby increasing the net income from unit holdings with the following component programmes.

Laying out of demonstration plots

Financial assistance of Rs.35,000 per ha in two annual installments is provided for adoption of integrated management practices in disease affected gardens. The objective of the programme is to encourage the farmers to adopt the technology developed for the management of disease affected coconut gardens. Demonstration plots will also help in motivate farmers in adoption of result oriented production technologies.

Integrated Farming for Productivity Improvement on cluster basis: The objective of the Coconut Clusters is increasing the production and productivity of coconut from unit holdings by proper and timely adoption of package of practices in a farmer participatory mode. This programme will facilitate the adoption of appropriate coconut based farming systems and promote farm level processing for value addition on a community basis. The scheme is being implemented by the Board on cluster basis in a contiguous area of appropriate size of 25-50 ha irrespective of the individual size of the holdings. The selection of the cluster is based on criteria such as demonstration value, easy accessibility, availability of minimum infrastructure facilities for the adoption of average management practices, cohesiveness of the group and most importantly the readiness of the farmers in the cluster to assume responsibility and implement the programme in a farmer participatory mode as per the Board's guidelines.

Assistance for Organic Manure Units

To promote the use of organic manure like vermicompost, coir pith compost, ordinary compost and FYM in coconut holdings. Financial assistance of Rs.20000 per unit or 50% of cost of production is provided for setting up of a unit.

Technology Demonstration

- Maintenance of Pilot Testing Plant and Quality Control laboratories.
- Techno-economic studies on product diversification and byproduct utilization
- Consultancy service on production, processing and marketing.
- Training programmes on convenience foods.

Market Promotion & Statistics

- Market information and intelligence service.
- Modernization of coconut processing by introduction of improved copra dryer/other processing machineries/ equipment @ 25% of the cost of dryer/equipment or maximum Rs.10,000, whichever is less.
- Surveys and evaluation studies.

Information and Information Technology

- Production of publications in different languages including grant-in-aid to State Government / Universities for production of publications.
- Production and screening of films including grant-in-aid to State Govt./Universities for production of films.
- Participation in exhibitions and fairs.
- Conducting seminars and workshops.
- Awards to best growers, master craftsmen, industrialists, processors and exporters.
- Training youths in harvesting and plant protection.
- Training farmers in scientific cultivation and post harvest processing.
- Training in the manufacture of coconut based handicrafts.
- National award to the best research worker or workers in product / machinery / equipment development.
- Establishment of National Information Centre cum Electronic Data Processing Unit.

Human Resource Development

- Training to improve technical skills of managerial and supervisory personnel.
- Training and visits to understand and appraise the technologies.

Financial Assistance under Technology Mission

Coconut Development Board invites applications for Financial Assistance under Technology Mission on Coconut. Applications are invited from prospective entrepreneurs for setting up / modernizing / expanding coconut processing units and promoting value added coconut based products.

Under Technology Mission on Coconut, financial assistance @ 25% of the project cost limited to a maximum of Rs.50.00 lakhs is extended as credit linked back ended subsidy for setting up of coconut processing units for production of value added coconut products. Modernization and expansion of existing processing units will also be eligible for assistance. The project for seeking assistance for setting up coconut processing unit should be submitted through the bank with a minimum of 40% of the project cost as term loan.

Packed, Branded coconut oil with Agmark standards, Virgin Coconut Oil (VCO), dietary fibre, Desiccated coconut powder, Coconut chips, defatted coconut powder, snow ball tender coconut, coconut milk powder, coconut jaggery, coconut cream, activated carbon, coconut water based vinegar, shell charcoal, packed tender nut water, minimally processed tender coconut, shell powder, shell/wood based handicrafts and copra dryer are some of the items which will be considered for assistance. Financial assistance will also be available for new and innovative coconut based products, which are backed by commercially viable technologies.

Financial assistance is provided for market promotion of coconut products through the media, participation in exhibitions / developing / hiring display facilities in super markets, adopting attractive packaging etc. @ 50% of the project cost limiting to Rs.10.00 lakhs for individuals and Rs.25.00 lakhs to co-Operative organisations.

Interested entrepreneurs can submit applications in the prescribed format along with project reports to the Chairman, Coconut Development Board, Kera Bhavan, Kochi 682 011.

Technology Mission on Coconut

Why the Mission?

Traditionally, coconut was grown for edible oil. It served as an ingredient for various industrial applications too. The changed food habits and availability of other cheaper edible oils both in the edible and industrial sectors, however, have brought out a drastic decline in the use of coconut oil in these areas. During the last few years, on account of heavy imports of cheaper vegetable oil, especially of the Palmolein, the price of coconut oil has been depressed despite the large-scale price support operations undertaken. The Price Support Scheme could not make much impact in pushing up the price level and was not beneficial to the farmers as expected. In this context, it was realized that diversification of coconut derived products and value addition could only help the coconut growers in getting remunerative prices. The coconut crop has also been affected by severe pests and debilitating diseases like root-wilt, ganoderma wilt, Thanjavur wilt and Tatipaka disease. It was realized that a major initiative should be started towards controlling the pests and diseases in coconut to improve its productivity and promote product diversification on better value realization from various coconut products. This will help the small and marginal farmers who depend on coconut for their livelihood to realize better returns. In this context to protect the interest of the coconut growers, the Prime Minister announced launching of Technology Mission on Coconut. The Mission should converge and synergize all the efforts through vertical and horizontal integration of existing programmes and address the problems and bridge the gaps through appropriate programmes in Mission Mode to ensure adequate, appropriate, timely and concurrent action. This would help develop a mechanism which makes coconut farming competitive and ensures reasonable returns.

Goals and Objectives

The **goals** and **objectives** of the Mission are to:

- *to establish* convergence and synergy among numerous ongoing governmental programmes in the field of coconut development in order to bring in horizontal and vertical integration of these programmes
- *to ensure* adequate, appropriate, timely and concurrent attention to all the links in the production, post harvest and consumption chain
- *to maximise* economic, ecological and social benefits from the existing investment and infrastructure created for coconut development
- *to promote* economically desirable diversification and value addition to generate skilled employment.
- *to disseminate* technologies using participatory approach through demonstration and promotion to address the gaps in a mission mode

Mission Approach

- mission approach is to evolve an approach for technology support which shall have synergy and convergence to address the existing gaps.
- existing schemes of Coconut Development Board and other institutes will continue with existing pattern and shall be converged in a manner that vertical and horizontal integration are achieved
- issues which have not been addressed in existing schemes to meet the challenges.
- issues relating to development of technologies for management of insect pests and disease affected gardens, product diversification and market promotion, its demonstration and promotion for adoption.

- missing links in existing programmes with focused attention to achieve the goals of the mission

Focus Areas

- Research and Development.
- Capacity building and participatory planning and implementation.
- Infrastructure Development.
- Integrated insect pest and disease management.
- Quality, Quantity and Productivity Improvement.
- Socially Acceptabie and Ecologically Sustainable Schemes leading to large scale adoption and long- lasting effects.
- Facilitating Credit Availability and Management.
- Poverty Alleviation Linked with Sustained Natural Resource
- Management through judicious management of existing land use.
- Post-harvest Processing, Product diversification and Value addition.
- Agri-Business in coconut.
- Equitable Access of People to benefits and equitable sharing by actually involving people at all levels during the project implementation, thus ensuring the evolution of a proper usufruct sharing mechanism for them.

Mission Components & Programmes

The Technology Mission covers four major components / programmes:

- Development and adoption of technologies for management of insect pests and disease affected coconut gardens
- Development and adoption of technologies for processing and product diversification.
- Market research and promotion
- Technical support, external evaluation and Emergent requirement.

■■■

Index